Mathematics for Rocket Science and Aerospace Engineering: Mastering the Rocket Equation and Propulsion Calculations with Python

Jamie Flux

https://www.linkedin.com/company/golden-dawn-engineering/

Contents

1 The Tsiolkovsky Rocket Equation   17
    Historical Context and Significance 17
    Derivation of the Equation 17
    Assumptions in Derivation 18
    Applications and Limitations 19
    Python Code Snippet 19

2 Newton's Second Law in Rocket Science   22
    Fundamental Principles of Newton's Second Law . . 22
    Rocket Thrust and Mass Flow 22
    Variable Mass Systems and Dynamics 23
    Trajectory Planning through Newtonian Mechanics . 23
    Compensating for Gravitational Forces 24
    Force Balance and Stability Analysis 24
    Numerical Integration of Rocket Motion 24
    Python Code Snippet 24

3 Conservation of Momentum   28
    Fundamental Principles of Momentum Conservation 28
    Application in Rocket Propulsion 28
    Implications for Effective Velocity Changes 29
    Ensuring Stability of Flights 29
    Numerical Methods for Momentum-Based Simulations 30
    Python Code Snippet 30

4 Mass Flow Rate Calculations   33
    Definition and Importance of Mass Flow Rate 33
    Relations to Thrust Generation 33
    Impact on Engine Performance 34
    Equations of State and Nozzle Flow 34

  Adjusting Performance Through Mass Flow Rate
  Modulation 35
  Python Code Snippet 35

5 Specific Impulse (Isp)   38
  Definition and Mathematical Representation 38
  Applications in Rocketry 38
  Comparative Analysis Across Propulsion Technologies 39
  1 Chemical Propulsion 39
  2 Electric Propulsion 39
  3 Hybrid Propulsion 39
  Theoretical Implications 40
  Integration into System Design 40
  Calculation Examples with Python Code 40
  Python Code Snippet 41

6 Thrust Equation for Chemical Rockets   43
  Thrust Equation Fundamentals 43
  Exhaust Velocity and Its Determination 43
  Chamber Pressure Effects 44
  Pressure Differential Contributions 44
  Implications for Propulsion Efficiency 44
  Python Code Snippet 45

7 Thrust Equation for Electric Propulsion   47
  Introduction to Electric Propulsion Systems 47
  Ion Thruster Thrust Fundamentals 47
  Exhaust Velocity in Ion Propulsion 48
  Hall Effect Thruster Dynamics 48
  Mass Flow Rate Calculations 48
  Ionization and Neutralization Efficiency 48
  Magnetized Plasma Interactions in Hall Effect Thrusters 49
  Thrust Characteristics and Optimization 49
  Energy and Power Considerations 49
  Python Code Snippet 50

8 Vacuum and Sea Level Thrust   53
  Introduction to Thrust Calculations 53
  Sea Level Thrust 53
  Vacuum Thrust 54
  Nozzle Design and Expansion Ratio 54
  Thrust Coefficient Analysis 54

Role of Specific Impulse 55
Implications for Launch Vehicle Design 55
Python Code Snippet 55

9 Delta-V Budgeting and Mission Planning 59
Delta-V and Its Significance in Space Missions . . . 59
Derivation and Interpretation of the Delta-V Budget 59
Allocation of Delta-V in Mission Phases 60
Delta-V Requirements for Common Orbital Transfers 60
Efficient Delta-V Utilization in Interplanetary Missions 60
Mission Planning Considerations and Constraints . . 61
Numerical Methods for Delta-V Optimization 61
Python Code Snippet 61

10 Orbital Mechanics Fundamentals 64
Kepler's Laws of Planetary Motion 64
1   First Law: The Law of Ellipses 64
2   Second Law: The Law of Equal Areas 64
3   Third Law: The Harmonic Law 65
Newton's Law of Universal Gravitation 65
Equation of Motion for Orbits 65
Vis-Viva Equation 66
Specific Orbital Energy 66
Orbital Elements Determination 66
Python Code Snippet 66

11 Hohmann Transfer Orbit 70
Introduction to Hohmann Transfer Orbits 70
Mathematical Framework 70
1     $\Delta V Calculations$ 70
2   Efficiency and Constraints 71
Application in Interplanetary Missions 72
Python Code Snippet 72

12 Bi-Elliptic Transfer 74
Conceptualization of Bi-Elliptic Transfer 74
Mathematical Derivation 74
1   Transfer Architecture 74
2   Velocity and Delta-V Calculations 75
Assessment of Efficiency in Transfers 76
Comparative Application in Interplanetary Missions 76

        Python Code Snippet 76

13 Patched Conic Approximations   78
        Overview of Patched Conic Approximation 78
        Framework of Patch Points 78
        1   Mathematical Formulation 79
        2   Conic Section Determination 79
        Transition Between Conics. 79
        1   Earth Departure to Heliocentric Orbit 79
        2   Heliocentric to Target Body Orbit 80
        Application of Patched Conics. 80
        1   Analytical Trajectory Simplification 80
        2   Planetary Flyby Calculations 80
        Python Code Snippet 81

14 Thrust-to-Weight Ratio   84
        Thrust-to-Weight Ratio Fundamentals 84
        Static Thrust-to-Weight Ratio. 84
        Thrust-to-Weight Ratio Dynamics During Ascent . . 85
        Payload Capacity and Thrust-to-Weight Ratio Optimization 85
        Applications in Different Phases of Flight 86
        Python Code Snippet 86

15 Engine Cycle Efficiency   90
        Efficiency Metrics in Rocket Engine Cycles 90
        Staged Combustion Cycle Efficiency 91
        Expander Cycle Efficiency 91
        Influence of Cycle Parameters on Efficiency 91
        Mathematical Modeling and Simulation 92
        Python Code Snippet 92

16 Pressure and Temperature Calculations   95
        Thermodynamic Principles in Rocket Engines 95
        Isentropic Flow and Engine Performance 95
        Pressure and Temperature in Nozzle Design 96
        Influence of Combustion on Pressure and Temperature 97
        Structural Material Considerations for Pressure and Temperature 97
        Python Code Snippet 98

17 Empirical Equations for Exhaust Velocity — 102
Introduction to Exhaust Velocity Calculations 102
Characteristic Velocity Consideration 102
Nozzle Efficiency and Design Implications 103
1   Empirical Correction Factors 103
Variable Propellant Combinations 103
1   Propellant Performance Indices 104
Conclusion on Exhaust Velocity Optimization 104
Python Code Snippet 104

18 Aerodynamics in Rocket Design — 107
Introduction to Aerodynamic Forces 107
Pressure Drag and Lift Components 107
Mathematical Modeling of Rocket Shape 108
Optimization Techniques in Rocket Aerodynamics . 108
1   Empirical Corrections and Fluid-Structure Interaction 108
Python Code Snippet 109

19 Drag Force Calculations — 112
Fundamentals of Drag Force 112
Components of Drag 112
1   Pressure Drag 112
2   Skin Friction Drag 113
3   Wave Drag 113
Advanced Computational Techniques 113
Minimization of Drag Forces 114
1   Streamlining and Shape Optimization 114
2   Re-entry Heating and Ablative Techniques . . 114
Empirical and Experimental Methods 114
Python Code Snippet 115

20 Rocket Stability and Control — 118
Stability Criteria and Equations 118
Mathematical Models for Control Systems 119
Guidance System Equations and Feedback Control . 119
Rotational Dynamics and Moment Equations 119
Trajectory Optimization and Stability Analysis . . . 120
Python Code Snippet 120

21 Quaternion Mathematics for Attitude Control — 124
- Introduction to Quaternions 124
- Quaternion Algebra and Operations 124
- Quaternion Representation of Rotations 125
- Quaternion Kinematics 125
- Application in Control Systems 125
- Integration in Guidance Systems 126
- Python Code Snippet 126

22 Gyroscopic Effects and Calculations — 129
- Fundamentals of Gyroscopic Motion 129
- Mathematical Model of Gyroscopic Stability 129
- Gyroscopic Precession and Nutation 130
- Application of Gyroscopic Principles in Flight Control 130
- Python Code Snippet 131

23 Calculating Structural Loads — 133
- Introduction to Structural Load Analysis 133
- Stress and Strain Fundamentals 133
- Load Types and Distribution 134
- Dynamic Loading Conditions 134
- Finite Element Analysis for Complex Structures . . 135
- Thermal Loads and Material Responses 135
- Load Factors and Safety Margins 135
- Python Code Snippet 136

24 Heat Shield Calculations — 140
- Heat Transfer Mechanisms 140
- Thermal Response Models 141
- Material Selection and Performance 141
- Thermo-mechanical Stress Analysis 141
- Validation against Empirical Data 142
- Future Considerations 142
- Python Code Snippet 143

25 Thermal Control Systems for Spacecraft — 147
- Heat Transfer in Spacecraft 147
- Thermal Control Mechanisms 148
- Active Thermal Regulation 148
- Passive Thermal Control Strategies 148
- Thermal Analysis Techniques 149
- Python Code Snippet 149

26 Escape Velocity and Gravity Assists — 153
Escape Velocity Concept 153
Gravitational Potential Energy and Kinetic Energy Balance 153
Trajectory Optimization for Gravity Assists 154
    1   Gravity Assist Mechanics 154
    2   Velocity Increment through Swingby 154
Application in Mission Design 155
    1   Patched Conic Approximations 155
Python Code Snippet 155

27 Non-Linear Dynamics of Rocket Systems — 158
Introduction to Non-Linear Dynamics 158
Equations of Motion 158
Aerodynamic Non-Linearities 159
Structural Flexibility 159
Propellant Slosh Dynamics 159
Control System Interactions 160
Numerical Methods for Non-Linear Systems 160
Applications of Non-Linear Analysis 160
Python Code Snippet 161

28 Time of Flight Calculations — 164
Introduction to Time of Flight 164
Fundamental Equations 164
Velocity and Acceleration Profiles 165
Trajectory Optimization Techniques 165
Numerical Integration for Time of Flight 165
Docking Sequence Timing 166
Case Study: Rendezvous Mission 166
Python Code Snippet 167

29 Fuel Optimal Control Techniques — 169
Introduction to Fuel Optimization 169
Optimal Control Formulations 169
Dynamic Equations and Constraints 170
Applications of Pontryagin's Maximum Principle . . 170
Fuel Optimal Thrust Arcs 170
Numerical Methods for Solving Optimal Control Problems 171
Case Study: Low Thrust Trajectory Optimization . 171
Python Code Snippet 171

30 Kalman Filter Applications in Navigation — 175
Introduction to Kalman Filters in Aerospace Navigation 175
Mathematical Foundations of the Kalman Filter . . 175
Prediction and Update Equations 176
Applications in Spacecraft Navigation 177
Extended and Unscented Kalman Filters 177
Implementation in Rocket Guidance Systems 177
Python Code Snippet 178

31 Monte Carlo Simulations in Propulsion — 180
Introduction to Monte Carlo Simulations 180
Fundamental Principles of Monte Carlo Methods . . 181
Application to Propulsion System Uncertainties . . . 181
Risk Assessment via Monte Carlo Simulations 182
Mathematical Implementation of Monte Carlo Methods 182
System Reliability Studies 182
Python Code Snippet 183

32 Benefits of Multi-Stage Rockets — 186
Introduction to Multi-Stage Rockets 186
Tsiolkovsky Rocket Equation for Multi-Stage Analysis 186
Payload Fraction Optimizations 187
Staging Strategies and Configuration Analysis 187
Velocity Increment Considerations and Stage Coupling 188
Advanced Multi-Stage System Design 188
Python Code Snippet 189

33 Geosynchronous and Polar Orbits — 192
Introduction to Orbital Dynamics 192
Geosynchronous Orbit Calculations 192
Station-Keeping in Geosynchronous Orbits 193
Polar Orbit Initial Mission Analysis 193
Sun-Synchronous Polar Orbits 194
Coverage Optimization in Polar Orbits 194
Configuration of Launch Elements for Polar Missions 194
Propellant Budgeting for Orbital Adjustments in Polar Regions 195
Python Code Snippet 195

34 Interplanetary Transfer Calculations — 198
- Foundations of Interplanetary Mechanics 198
- Hohmann Transfer Calculations 198
- Bi-Elliptic Transfer Considerations 199
- Influence of Gravitational Assists 199
- Patched Conic Approximations 200
- N-Body Problem and Numerical Methods 200
- Optimization of Trajectory Pathways 201
- Python Code Snippet 201

35 Ion Propulsion System Calculations — 204
- Fundamentals of Ion Propulsion 204
- Ionization and Acceleration of Propellant 204
- Thrust Power Calculations 205
- Grid Erosion and System Longevity 205
- Optimization Techniques for Deep Space Missions . 206
- Mathematical Modeling in Ion Thruster Design . . . 206
- Python Code Snippet 207

36 Hybrid Rocket Propulsion — 210
- Introduction to Hybrid Rocket Propulsion 210
- Combustion Processes in Hybrid Rockets 210
- Solid Fuel Regression Rates 211
- Thermodynamic Analysis 211
- Design Considerations for Hybrid Systems 212
- Hybrid Rocket Propellant Materials 212
- System Integration Principles 213
- Python Code Snippet 213

37 Rocket Propellant Thermochemistry — 216
- Fundamental Thermochemistry of Rocket Propellants 216
- Heat of Combustion Calculations 216
- Chemical Equilibrium and Thermodynamic Efficiency 217
- Adiabatic Flame Temperature 217
- Thermochemical Reaction Kinetics 217
- Energy Efficiency Metrics 218
- Python Code Snippet 218

38 Combustion Efficiency Evaluation — 222
- Introduction to Combustion Efficiency 222
- Chemical Reaction Modeling 222
- Heat Release and Energy Conversion 223

Efficiency Metrics and Performance	223
Combustion Kinetics	223
Thermochemical Efficiency Optimization	224
Python Code Snippet	224

39 Nozzle Design and Optimization 228
Fundamentals of Nozzle Theory	228
The Role of Mach Number in Nozzle Design	228
Isentropic Flow Relationships	229
Underexpanded and Overexpanded Nozzles	229
Thrust Optimization Techniques	230
Thermodynamic Considerations in Nozzle Design	230
Materials and Structural Integrity	230
Python Code Snippet	231

40 Rocket Reusability Models 234
Economic Implications of Reusability	234
Structural Design for Reusability	234
Thermal Protection and Recovery Systems	235
Dynamics and Control Systems for Reusable Rockets	235
Cost-Benefit Analysis of Reusability	235
Safety Considerations in Multiple Flight Cycles	236
Python Code Snippet	236

41 Stage Separation Dynamics 240
Dynamics of Stage Separation	240
Separation Timing and Trajectory Accuracy	241
Structural Loads During Separation	241
Safety Protocols in Stage Separation	241
Control Systems and Stage Separation	242
Empirical and Computational Modeling	242
Python Code Snippet	242

42 Propellant Mixing and Flow Dynamics 246
Kinematics of Propellant Mixing	246
Shear Layer Formation	246
Mixing Efficacy in Nozzles	247
Influence of Turbulence on Mixing	247
Heat Transfer Influences on Mixing	248
Empirical Modeling of Mixing Processes	248
Applications of Computational Fluid Dynamics in Mixing Analysis	248

      Python Code Snippet 249

43 Propagation of Errors in Trajectory — 252
      Error Sources in Trajectory Planning 252
      Mathematical Representation of Error Propagation . 252
      Covariance Analysis for Trajectory Errors 253
      Monte Carlo Simulations 253
      Optimization Techniques to Minimize Errors 253
      Robust Control Implementation 254
      Application of Kalman Filters in Error Propagation 254
      Python Code Snippet 254

44 Surface Roughness in Heat Exchange — 258
      Introduction to Surface Roughness in Aerospace Systems 258
      Quantifying Surface Roughness 258
      Impact of Surface Roughness on Heat Transfer Coefficients 259
      Pressure Drop Considerations 259
      Design Optimization in Rocket Engine Components 260
      Empirical Correlations for Aerospace Applications . 260
      Python Code Snippet 260

45 Material Fatigue Analysis — 264
      Introduction to Fatigue in Aerospace Materials . . . 264
      Fundamental Concepts of Material Fatigue 264
      Cyclic Load and Microstructural Effects 265
      Paris' Law for Crack Growth Prediction 265
      Miner's Rule for Cumulative Damage 265
      Computational Fatigue Analysis Techniques 266
      Fatigue Testing and Validation 266
      Python Code Snippet 266

46 Supersonic Flow Calculations — 269
      Introduction to Supersonic Flow 269
      Fundamental Equations of Supersonic Flow 269
      Shock Wave Formation 270
      Drag Reduction Techniques 270
      Numerical Methods for Supersonic Flow 271
      Applications in Modern Rocket Design 271
      Python Code Snippet 271

47 Flow Separation and Boundary Layer Theory — 274
Boundary Layer Formation in High-Speed Flow . . . 274
Turbulent Boundary Layer Considerations 274
Flow Separation Mechanisms 275
Calculations of Displacement and Momentum Thickness 275
Impact of Compressibility on Boundary Layer Development 276
Mitigation Strategies for Flow Separation 276
Python Code Snippet 276

48 Free-Molecular Flow and its Effects — 280
Regime Characteristics of Free-Molecular Flow . . . 280
Mathematical Modeling of Free-Molecular Flow . . . 280
Drag Force in Free-Molecular Flow 281
Thermal Effects and Energy Transfer 281
Predicting Vehicle Performance in Free-Molecular Flow 281
Python Code Snippet 282

49 Cryogenic Propellant Management — 285
Introduction to Cryogenic Propellants 285
Thermodynamic Properties of Cryogenic Fluids . . . 285
Heat Transfer Mechanisms in Cryogenic Storage . . 286
Cryogenic Fluid Dynamics 286
Boil-off and Phase Change Calculations 286
Storage Tank Design Considerations 287
Conclusion 287
Python Code Snippet 287

50 Forced and Natural Convection in Rockets — 290
Fundamentals of Convection Mechanisms 290
Mathematical Modeling of Forced Convection 290
Natural Convection in Rocket Systems 291
Applications in Rocket Thermal Management 292
Significance of Convection Studies in Rocketry . . . 292
Python Code Snippet 292

51 Failure Mode and Effect Analysis (FMEA) — 295
Introduction to FMEA in Propulsion Systems 295
FMEA Methodology 295
    1   Severity Assessment 296

    2   Occurrence Evaluation 296
    3   Detection Quality 296
    Data-Driven Approaches in FMEA 296
    1   Machine Learning in FMEA 296
    2   Probabilistic Risk Assessment 297
    Implementation in Aerospace Propulsion Systems . . 297
    Python Code Snippet 297

52 Gas Dynamics in Variable Cross-Sections   301
    Fundamental Principles of Gas Dynamics 301
    Application of Bernoulli's Equation 301
    Isentropic Flow Relations 302
    1   Mach Number Variation 302
    Shock Waves and Expansion Fans. 302
    Nozzle Design and Optimization 303
    1   Design Considerations 303
    Python Code Snippet 304

53 Boundary Conditions in Computational Fluid Dynamics (CFD)   307
    Introduction to Boundary Conditions in CFD 307
    Types of Boundary Conditions 307
    1   Dirichlet Boundary Condition 307
    2   Neumann Boundary Condition 308
    3   Robin Boundary Condition 308
    4   Periodic Boundary Condition 308
    Implementation in CFD Solvers 309
    Challenges and Considerations 309
    1   Alignment to Physical Laws 309
    2   Incompatible Domain Configurations 309
    3   Numerical Stability and Convergence 310
    Python Code Snippet 310

54 Vortex Dynamics in Rocket Exhaust   313
    Fundamentals of Vortex Dynamics 313
    Vortex Formation in Rocket Exhaust 313
    Mathematical Modeling of Vortex Dynamics 314
    Techniques for Mitigating Vortex-Induced Disturbances 314
    Numerical and Computational Approaches 315
    Python Code Snippet 315

55 Dynamic Stability Margin Calculations — 318
    Dynamic Stability Fundamentals 318
    Defining Stability Margins 319
    Computing Dynamic Stability Margins 319
    Techniques for Margin Enhancement 320
    Example Application 320
    Python Code Snippet 320

56 Advanced Finite Element Analysis (FEA) — 324
    Introduction to Finite Element Analysis 324
    Formulation of Element Stiffness Matrix 325
    Meshing Techniques in Aerospace Applications . . . 325
    Non-linear FEA Challenges 325
    Dynamic Analysis in FEA 326
    Applications of FEA in Aerospace Design 326
    Python Code Snippet 327

57 Micro Thrusters and Satellite Attitude Control — 330
    Fundamentals of Attitude Control using Micro Thrusters 330
    Micro Thruster Types and Their Efficiency 331
    Propellant Mass and Thruster Sizing 331
    Control Strategies in Attitude Modulation 331
    Mathematical Modeling of Micro Thruster Dynamics 332
    Challenges and Optimization of Micro Thruster Systems 332
    Python Code Snippet 333

58 Boundary Layer Control in Rockets — 337
    Introduction to Boundary Layer Theory 337
    Techniques for Boundary Layer Control 337
    1   Surface Treatment and Shaping 338
    2   Boundary Layer Suction and Blowing 338
    Impact on Rocket Performance 338
    Python Code Snippet 339

59 Spacecraft Rendezvous and Docking Algorithms — 342
    Rendezvous Basics 342
    Clohessy-Wiltshire Equations 342
    Phasing Maneuvers 343
    Docking Dynamics 343
    Optimal Control in Rendezvous 343
    Python Code Snippet 344

60 Leaky Integrations in Control Systems — 348
- Introduction to Leaky Integrators 348
- Feedback Loop Design with Leaky Integrators 348
- Stability Analysis Through Eigenvalue Evaluation . 349
- Implementation in Rocket Attitude Control Systems 349
- Sensitivity Functions and Robustness 350
- Python Code Snippet 350

61 Limitations of Linear Control Techniques — 354
- Linear Control Approaches in Aerospace Engineering 354
- Challenges in Dynamic Rocket Systems 354
- Stability and Robustness Limitations 355
- Transition to Non-Linear Control Methods 355
- Model Predictive Control and Adaptive Techniques . 356
- Python Code Snippet 356

62 Genetic Algorithms for Optimizing Propulsion — 359
- Introduction to Genetic Algorithms 359
- Chromosome Representation and Fitness Evaluation 359
- Selection Mechanisms and Genetic Operators 360
- Convergence and Computational Considerations . . 360
- Application in Propulsion System Design 361
- Python Code Snippet 361

63 Data Fusion in Multi-Sensor Navigation Systems — 365
- Introduction to Multi-Sensor Data Fusion 365
- Kalman Filter for Data Fusion 365
- Nonlinear Filtering Techniques 366
- Implementing Data Fusion for Enhanced Navigation Accuracy 367
- Python Code Snippet 367

64 Sounding Rockets and High-Altitude Research — 370
- Introduction 370
- Launch Dynamics and Altitude Achievements 370
- 1 Payload Considerations 371
- Trajectory and Guidance Systems 371
- Aerodynamic and Thermal Constraints 372
- 1 Material Selection and Fabrication 372
- Data Collection and Analysis 372
- Python Code Snippet 373

65 Space Weather Effects on Rocket Launch — 376
    Introduction to Space Weather Phenomena 376
    Electromagnetic Interference and Mitigation 376
    Radiation Effects on Electronics 377
    Atmospheric Density Variations and Trajectory Impact 377
    Thermal and Structural Considerations 378
    1   Material Selection and Stress Analysis 378
    Mitigation Strategies for Launch Environment 378
    Python Code Snippet 378

66 Astrodynamics of Hypersonic Vehicles — 382
    Introduction to Hypersonic Regimes 382
    Atmospheric Re-entry Dynamics 382
    Control System Response 383
    Thermal Protection Systems 383
    Propulsion Systems Integration 383
    Navigation in Hypersonic Flight 384
    Mission Design Considerations 384
    Python Code Snippet 384

Chapter 1

The Tsiolkovsky Rocket Equation

The Tsiolkovsky Rocket Equation, often hailed as the bedrock of rocket dynamics, establishes a fundamental relationship between the change in velocity of a rocket and the mass of the propellant expended. As a seminal contribution by Konstantin Tsiolkovsky, this equation is indispensable in understanding the principles of rocket propulsion.

Historical Context and Significance

The derivation of the Tsiolkovsky Rocket Equation dates back to the late 19th century, wherein Tsiolkovsky, a pioneering Russian scientist, laid the groundwork for modern astronautics. This equation provided the first quantitative understanding of how velocity change in a rocket is influenced by expelling mass, paving the way for subsequent advances in space exploration.

Derivation of the Equation

The derivation begins with the application of the law of conservation of momentum. Consider a rocket of initial mass m_0 (including propellant) traveling at velocity v. When the rocket expels a small mass Δm at velocity v_e relative to the rocket, the change in momentum must equate to zero for an isolated system:

$$(m - \Delta m)(v + \Delta v) + \Delta m(v - v_e) = mv \tag{1.1}$$

Neglecting terms with $\Delta m \Delta v$ and simplifying, the following differential equation is obtained:

$$m\,\Delta v = v_e\,\Delta m \tag{1.2}$$

Rearranging the terms yields:

$$\Delta v = v_e \frac{\Delta m}{m} \tag{1.3}$$

Integrating both sides with respect to their respective variables provides the fundamental form of the Tsiolkovsky Rocket Equation:

$$\Delta v = v_e \ln\left(\frac{m_0}{m_f}\right) \tag{1.4}$$

where Δv is the change in velocity, v_e is the effective exhaust velocity, m_0 is the initial total mass, and m_f is the final total mass.

Assumptions in Derivation

Key assumptions inherent in the derivation are crucial for understanding the applicability and limitations of the Tsiolkovsky Rocket Equation:

1. **Isolated System:** The equation assumes that external forces, such as gravity and drag, are negligible. It applies most accurately to the rocket in the vacuum of space and less so in atmospheric conditions.

2. **Steady State:** The effective exhaust velocity v_e is considered constant, assuming an idealized system where propellant gases are expelled uniformly.

3. **Rigid Body Dynamics:** The equation does not account for changes in the rocket structure or configuration during propellant burn.

4. **No Rotational Motion:** Any rotational dynamics or forces acting upon them are neglected, focusing solely on translational motion.

Applications and Limitations

The Tsiolkovsky Rocket Equation is instrumental in the design and analysis of both single-stage and multi-stage rockets. However, its practical application requires modification and extension to account for non-ideal factors in real-world scenarios:

- **Gravity Losses:** Adjustments are necessary to account for additional delta-v required to overcome gravitational pull during an ascent in planetary bodies' atmospheres.
- **Aerodynamic Drag:** Realistic models incorporate losses due to atmospheric drag, particularly during launch and ascent phases.
- **Variable Exhaust Velocity:** Some advanced propulsion systems may experience variations in exhaust speed, for which further analysis is warranted, often involving empirical measurement and computational models.

In summary, while the Tsiolkovsky Rocket Equation provides a fundamental starting point for propulsion calculations, it requires careful consideration of external factors and system specifics when applied to practical engineering problems.

Python Code Snippet

Below is a Python code snippet that encapsulates the algorithms and calculations presented in this chapter, providing computational tools to derive the Tsiolkovsky Rocket Equation and apply it effectively for various parameters.

```
import numpy as np

def tsiolkovsky_rocket_equation(m0, mf, ve):
    '''
    Compute the change in velocity (Delta-v) using the Tsiolkovsky
    ↪ Rocket Equation.
    :param m0: Initial total mass (including propellant).
    :param mf: Final total mass (after propellant is burnt).
    :param ve: Effective exhaust velocity.
    :return: Change in velocity (Delta-v).
    '''
    return ve * np.log(m0 / mf)

def rocket_equation_with_gravity_drag(m0, mf, ve, gravity_losses,
    ↪ drag_losses):
    '''
```

```python
    Adjust the Tsiolkovsky Rocket Equation to account for gravity
    ↪ and drag losses.
    :param m0: Initial total mass.
    :param mf: Final total mass.
    :param ve: Effective exhaust velocity.
    :param gravity_losses: Additional delta-v needed to overcome
    ↪ gravity.
    :param drag_losses: Additional delta-v needed to overcome drag.
    :return: Effective change in velocity.
    '''
    delta_v = ve * np.log(m0 / mf)
    return delta_v - gravity_losses - drag_losses

def calculate_mass_ratio(m0, delta_v, ve):
    '''
    Calculate the mass ratio needed to achieve a given change in
    ↪ velocity.
    :param m0: Initial total mass.
    :param delta_v: Desired change in velocity.
    :param ve: Effective exhaust velocity.
    :return: Final mass (mf).
    '''
    mass_ratio = m0 / np.exp(delta_v / ve)
    return mass_ratio

# Example parameters
m0 = 5000   # Initial mass in kg
mf = 3000   # Final mass in kg
ve = 2500   # Exhaust velocity in m/s
gravity_losses = 500   # Gravity losses in m/s
drag_losses = 200   # Drag losses in m/s

# Running calculations
delta_v = tsiolkovsky_rocket_equation(m0, mf, ve)
delta_v_corrected = rocket_equation_with_gravity_drag(m0, mf, ve,
↪ gravity_losses, drag_losses)
required_mass_ratio = calculate_mass_ratio(m0, delta_v, ve)

print("Calculated Delta-v:", delta_v, "m/s")
print("Corrected Delta-v including gravity and drag losses:",
↪ delta_v_corrected, "m/s")
print("Final mass required for the given Delta-v:",
↪ required_mass_ratio, "kg")
```

This code provides functions crucial for applying the Tsiolkovsky Rocket Equation to real-world rocket propulsion scenarios:

- `tsiolkosky_rocket_equation` function calculates the change in velocity using the foundational rocket equation for given mass values and exhaust velocity.

- `rocket_equation_with_gravity_drag` modifies the basic rocket equation to include additional delta-v for gravity and drag losses, representing a more realistic flight scenario.

- `calculate_mass_ratio` determines the mass ratio required based on a specified delta-v, allowing for planning of propellant loads.

The example calculations illustrate how to apply these mathematical constructs for typical rocket mission planning.

Chapter 2

Newton's Second Law in Rocket Science

Fundamental Principles of Newton's Second Law

Newton's Second Law is foundational to the study of dynamics, and its application in rocketry is pivotal for understanding how forces interact to produce motion. The law is mathematically represented as:

$$\mathbf{F} = m\mathbf{a} \qquad (2.1)$$

where $\mathbf{F}$ is the force acting on an object, m is the mass of the object, and $\mathbf{a}$ is the acceleration. In the context of a rocket, the law describes how the mass of the rocket and the forces applied to it govern its acceleration.

Rocket Thrust and Mass Flow

Thrust (T), the force propelling a rocket, is derived from the expulsion of mass at high speed. According to Newton's Second Law, the thrust equation is articulated as:

$$T = \dot{m} v_e \qquad (2.2)$$

where $\dot{m}$ is the mass flow rate of the expelling gases, and v_e is the effective exhaust velocity. This equation captures how rapid expulsion of mass accelerates the rocket forward, with thrust being directly proportional to both the mass flow rate and the effective exhaust velocity.

Variable Mass Systems and Dynamics

Rockets exemplify variable mass systems, where the mass decreases as propellant is consumed. The dynamic equation for such systems extends Newton's Second Law:

$$\mathbf{F} = \frac{d(m\mathbf{v})}{dt} \tag{2.3}$$

Considering the mass variation, the expanded version becomes:

$$\mathbf{F} = m\frac{d\mathbf{v}}{dt} + \mathbf{v}\frac{dm}{dt} \tag{2.4}$$

For a rocket with no external forces, the thrust T equates to the negative of the second term, leading to $T = \mathbf{v}\frac{dm}{dt}$. This highlights the significance of exhaust velocity and mass rate change in determining thrust.

Trajectory Planning through Newtonian Mechanics

In the design and execution of flight paths, Newton's Second Law aids in trajectory planning by correlating force and acceleration with path alterations. Assuming net external force $\mathbf{F}_{\text{net}}$, the velocity and position vectors $\mathbf{v}$ and $\mathbf{r}$ are computed through:

$$\mathbf{a} = \frac{\mathbf{F}_{\text{net}}}{m} \tag{2.5}$$

Hence, trajectory is integrated from acceleration:

$$\mathbf{v}(t) = \int \mathbf{a}(t)\,dt + \mathbf{v}_0 \tag{2.6}$$

$$\mathbf{r}(t) = \int \mathbf{v}(t)\,dt + \mathbf{r}_0 \tag{2.7}$$

Here, $\mathbf{v}_0$ and $\mathbf{r}_0$ represent the initial conditions of velocity and position, respectively. The exact path taken by a rocket is thus determined by the integration of the changing forces and resultant accelerations over time.

Compensating for Gravitational Forces

The influence of gravity demands compensation, modifying the net force:

$$\mathbf{F}_{\text{net}} = T - m\mathbf{g} \tag{2.8}$$

where $\mathbf{g}$ is the gravitational acceleration vector. Incorporating this into the trajectory planning ensures that the rocket achieves the desired ascent despite gravitational pull.

Force Balance and Stability Analysis

The balance of forces, including those due to thrust, gravity, and aerodynamic drag (D), determines stability:

$$\mathbf{F}_{\text{net}} = T - m\mathbf{g} - D \tag{2.9}$$

Stability analysis requires assessing how these forces interplay during various flight phases to maintain desired orientations and prevent undesired motions or deviations from the intended flight path.

Numerical Integration of Rocket Motion

Numerical methods, such as finite-difference techniques, facilitate the integration of motion equations to predict trajectories. Implementing algorithms in computational tools allows for accurate simulation of flight dynamics over discrete time steps, ensuring that force applications approximate continuous changes in motion dynamics.

Python Code Snippet

Below is a Python code snippet that encompasses the core computational elements relevant to Newton's Second Law applications in rocket science, including the calculation of force, thrust, trajectory planning, and numerical integration.

```python
import numpy as np
from scipy.integrate import odeint

def force(mass, acceleration):
    '''
    Calculate the force using Newton's Second Law.
    :param mass: Mass of the object.
    :param acceleration: Acceleration of the object.
    :return: Force acting on the object.
    '''
    return mass * acceleration

def thrust(mass_flow_rate, exhaust_velocity):
    '''
    Calculate the thrust produced by a rocket engine.
    :param mass_flow_rate: Mass flow rate of expelled gases.
    :param exhaust_velocity: Effective exhaust velocity.
    :return: Thrust produced by the engine.
    '''
    return mass_flow_rate * exhaust_velocity

def trajectory(mass, force_net, v0, r0, t):
    '''
    Calculate the velocity and position vectors from given forces.
    :param mass: Mass of the object.
    :param force_net: Net external force.
    :param v0: Initial velocity vector.
    :param r0: Initial position vector.
    :param t: Time array for integration.
    :return: Computed velocity and position vectors.
    '''
    def motion_equations(y, t, mass, force_net):
        v, r = y[:3], y[3:]
        a = force_net / mass  # Acceleration
        dydt = np.concatenate((a, v))
        return dydt

    y0 = np.concatenate((v0, r0))
    sol = odeint(motion_equations, y0, t, args=(mass, force_net))
    return sol[:, :3], sol[:, 3:]  # velocity, position

def compensate_gravity(thrust, mass, gravity, drag):
    '''
    Compensate for gravitational and drag forces to find net thrust.
    :param thrust: Thrust force.
    :param mass: Mass of the object.
    :param gravity: Gravitational acceleration vector.
    :param drag: Aerodynamic drag force.
    :return: Net thrust force.
    '''
```

```python
    '''
    gravity_force = mass * gravity
    net_thrust = thrust - gravity_force - drag
    return net_thrust

# Constants and initial conditions
g = np.array([0, 0, -9.81])  # Gravity vector in m/s^2
mass_initial = 1000.0  # Initial mass in kg
mass_flow_rate = 5.0  # Mass flow rate in kg/s
exhaust_velocity = 3000.0  # Exhaust velocity in m/s
drag_force = np.array([0.0, 0.0, -50.0])  # Drag force acting on
    rocket

# Calculate thrust
T = thrust(mass_flow_rate, exhaust_velocity)

# Compensate for gravity
net_force = compensate_gravity(T, mass_initial, g, drag_force)

# Initial velocity and position
v0 = np.array([0, 0, 0])  # Initial velocity in m/s
r0 = np.array([0, 0, 0])  # Initial position in meters

# Time array for simulation
time = np.linspace(0, 100, 500)  # 100 seconds, 500 steps

# Trajectory computation
velocity, position = trajectory(mass_initial, net_force, v0, r0,
    time)

# Display results
print("Thrust:", T)
print("Net Force:", net_force)
print("Final Velocity:", velocity[-1])
print("Final Position:", position[-1])
```

This code defines several key functions essential for understanding Newton's Second Law in the context of rocketry:

- `force` function calculates the force exerted on a rocket using its mass and acceleration.

- `thrust` computes the thrust produced by the rocket based on the mass flow rate and exhaust velocity.

- `trajectory` integrates the equations of motion to predict the rocket's velocity and position over time.

- `compensate_gravity` accounts for gravitational and drag forces to determine the net force acting on the rocket.

The final block of code provides an example calculation showing the thrust, net force, and predicted velocity and position of the rocket over a simulated flight time.

Chapter 3

Conservation of Momentum

Fundamental Principles of Momentum Conservation

In the theoretical framework of classical mechanics, momentum conservation serves as a critical paradigm, crucially underpinning the mechanics of rocket propulsion. The principle of conservation of momentum is predicated on the notion that a closed system's total momentum remains invariant over time in the absence of external forces. The total linear momentum **P** of a system comprising particles is mathematically encapsulated by:

$$\mathbf{P} = \sum_i m_i \mathbf{v}_i \tag{3.1}$$

where m_i and $\mathbf{v}_i$ represent the mass and velocity vectors of the ith particle, respectively. In the landscape of aerospace engineering, this foundational principle is instrumental in elucidating the mechanics of rocket propulsion.

Application in Rocket Propulsion

The utilization of the conservation of momentum principle in rocket propulsion enables the formulation of the rocket equation, revealing the mechanism by which rockets achieve acceleration in a vacuum.

The essential proposition is that the momentum lost by the expelled mass of propellant is counterbalanced by the momentum gained by the rocket. This balance is expressed as:

$$m\frac{d\mathbf{v}}{dt} = -\dot{m}\mathbf{v}_e \tag{3.2}$$

where m represents the mass of the rocket, $\mathbf{v}_e$ is the effective exhaust velocity, and $\dot{m}$ is the mass flow rate of the propellant. Conservation dictates that the rate of momentum change of the rocket is equivalent and opposite to that of the ejected gases.

Implications for Effective Velocity Changes

The propulsion of rockets, driven by momentum exchange, mandates precise manipulation of the velocity vector to achieve desired trajectory and velocity changes. The rocket's velocity increment, $\Delta\mathbf{v}$, is derived from the integration of the differential form of the rocket equation:

$$\Delta\mathbf{v} = \mathbf{v}_e \ln\left(\frac{m_0}{m_f}\right) \tag{3.3}$$

where m_0 and m_f are the initial and final total masses, respectively. This expression unveils the logarithmic dependency of velocity change on the mass ratio, accentuating the significance of exhaust velocity in augmenting $\Delta\mathbf{v}$.

Ensuring Stability of Flights

Momentum conservation is inherently linked to flight stability, as it dictates the symmetrical distribution of forces and thus influences the dynamic stability of the vehicle. Analyzing perturbations involves linearizing the equations of motion and examining the eigenvalues of the resulting system to ascertain dynamic stability criteria. Situational stability is achieved by maintaining consistent mass distribution, appropriate thrust alignment, and structural integrity, ensuring that the conservation principles hold valid under various flight regimes.

The forces and moments acting upon a rocket during its ascent can be expressed in terms of the derivatives of momentum, enabling the assessment and subsequent tuning of stability parameters:

$$\frac{d\mathbf{P}}{dt} = \mathbf{F}_{\text{thrust}} + \mathbf{F}_{\text{aero}} - m\mathbf{g} + \mathbf{F}_{\text{control}} \qquad (3.4)$$

where $\mathbf{F}_{\text{thrust}}$, $\mathbf{F}_{\text{aero}}$, $-m\mathbf{g}$, and $\mathbf{F}_{\text{control}}$ denote the thrust, aerodynamic, gravitational, and control forces, respectively.

Numerical Methods for Momentum-Based Simulations

Accurate simulation of rocket dynamics necessitates rigorous numerical techniques, such as the application of finite-difference schemes or Runge-Kutta methods for solving the differential momentum equations. Computational fluid dynamics (CFD) simulations further incorporate momentum conservation to predict aerodynamic effects, optimizing rocket design for enhanced performance and stability.

The discretization of temporal terms in the momentum equation provides insights into the evolutionary phase-space behavior of the rocket system, facilitated by algorithm implementations for momentum calculus in computational tools. Adjustments to momentum-based variables, such as mass flow rates and thrust alignments, permit strategic trajectory control, embodying the engineering prowess needed for successful mission profiles.

This text, structured in the formal style expected of an aerospace engineering doctoral thesis, emphasizes the precise and technical nature of momentum conservation and its application in the domain of rocketry. Each section focuses on unpacking key elements of the principle and its consequential role in ensuring effective aerospace operation and design.

Python Code Snippet

Below is a Python code snippet that encompasses the core computational elements of rocket propulsion calculations including the conservation of momentum equations, delta-v computations, stability assessments, and numerical simulations.

```
import numpy as np
from scipy.integrate import odeint
```

```python
def conservation_of_momentum(mass, v_exhaust, mass_flow_rate,
    delta_time):
    '''
    Calculate the change in velocity due to thrust using
        conservation of momentum.
    :param mass: Current mass of the rocket.
    :param v_exhaust: Effective exhaust velocity.
    :param mass_flow_rate: Rate of mass ejection (propellant).
    :param delta_time: Time increment for the calculation.
    :return: Increment in velocity.
    '''
    return v_exhaust * np.log(mass / (mass - mass_flow_rate *
        delta_time))

def rocket_equation(mass, v_exhaust, mass_initial, epsilon):
    '''
    Compute the total velocity change (delta-v) of a rocket.
    :param mass: Final mass of the rocket after burn.
    :param v_exhaust: Effective exhaust velocity of the rocket.
    :param mass_initial: Initial mass including full propellant
        load.
    :param epsilon: Terminal condition for convergence in iterative
        solutions.
    :return: Total delta-v achieved by the rocket.
    '''
    delta_v = 0
    current_mass = mass_initial
    while current_mass > mass + epsilon:
        delta_v += conservation_of_momentum(current_mass, v_exhaust,
            0.1, 0.1)
        current_mass -= 0.1  # Simulating mass flow

    return delta_v

def stability_assessment(p_initial, thrust, aero_forces, gravity,
    control_forces, time_span):
    '''
    Simulate rocket stability over a given time span considering
        forces acting on the vehicle.
    :param p_initial: Initial state (velocity and orientation).
    :param thrust: Thrust force vector.
    :param aero_forces: Aerodynamic forces vector.
    :param gravity: Gravity force vector.
    :param control_forces: Control input forces vector.
    :param time_span: Time span for the simulation.
    :return: Trajectories for position, velocity, and orientation.
    '''
    def equations_of_motion(state, t):
        v, o = state
        total_force = thrust + aero_forces - gravity +
            control_forces
        accel = total_force / p_initial['mass']
        return [accel, o]  # Simplified for demonstration purposes
```

```
    return odeint(equations_of_motion, p_initial['state'],
    ↪    time_span)

# Example of usage
mass_initial = 2000.0
mass_final = 1000.0
v_exhaust = 3000.0
epsilon = 1.0e-6

delta_v = rocket_equation(mass_final, v_exhaust, mass_initial,
↪    epsilon)
print("Total Delta-V:", delta_v)

# Stability simulation setup
initial_state = {'state': [0, 1], 'mass': mass_initial}
thrust = np.array([1000, 0])
aero_forces = np.array([-10, 2])
gravity = np.array([0, -9.81])
control_forces = np.array([5, 0])
time_span = np.linspace(0, 50, 500)

trajectories = stability_assessment(initial_state, thrust,
↪    aero_forces, gravity, control_forces, time_span)

print("Simulation completed, trajectories calculated.")
```

This code defines several key functions necessary for implementing rocket propulsion dynamics:

- `conservation_of_momentum` calculates the change in velocity as thrust is applied, based on the decrease in rocket mass over time.

- `rocket_equation` implements the iterative computation of total velocity change (Δv) as a function of mass ratio and exhaust velocity.

- `stability_assessment` simulates the rocket's stability and trajectory by solving the motion equations governed by contributing aerodynamic, gravitational, thrust, and control forces.

The final block of code demonstrates how these computational functions can be employed in practical scenarios for planning and analyzing rocket trajectories.

Chapter 4

Mass Flow Rate Calculations

Definition and Importance of Mass Flow Rate

In the context of rocket propulsion systems, mass flow rate ($\dot{m}$) is a crucial parameter that dictates thrust generation and overall engine performance. The mass flow rate is defined as the amount of mass expended by the engine per unit of time and is typically expressed in units of kilograms per second (kg/s). Mathematically, it can be described by:

$$\dot{m} = \frac{dm}{dt} \tag{4.1}$$

where dm is the differential element of mass and dt is the differential element of time. This parameter influences how efficiently a rocket can convert stored chemical energy into kinetic energy, subsequently affecting the vehicle's acceleration capabilities.

Relations to Thrust Generation

The thrust (F) produced by a rocket engine is fundamentally linked to the mass flow rate of its propellant. It is governed by the equation:

$$F = \dot{m} \cdot v_e + A \cdot (p_e - p_0) \qquad (4.2)$$

where v_e is the effective exhaust velocity, A is the area of the exhaust nozzle, p_e is the pressure at the nozzle exit, and p_0 is the ambient pressure. The term $\dot{m} \cdot v_e$ represents the momentum thrust, while $A \cdot (p_e - p_0)$ denotes the pressure thrust.

Impact on Engine Performance

High mass flow rates generally lead to greater thrust, assuming constant effective exhaust velocity. However, increased mass flow rates necessitate enhanced thermal management and structural integrity to withstand the thermal and mechanical stresses imposed on the engine components. Optimizing mass flow rate involves a balance between achieving sufficient thrust and maintaining engine performance parameters, including specific impulse (Isp), defined as:

$$I_{sp} = \frac{F}{\dot{m} \cdot g_0} \qquad (4.3)$$

where g_0 is the standard acceleration due to gravity at sea level.

Equations of State and Nozzle Flow

The propellant behavior in a rocket engine is described by the equations of state, specifically when expanding through the engine nozzle. The relationship between mass flow rate and chamber conditions is expressed as:

$$\dot{m} = \rho_c \cdot A_c \cdot v_c \qquad (4.4)$$

where ρ_c is the propellant density at the combustion chamber conditions, A_c is the chamber cross-sectional area, and v_c is the velocity of the propellant in the chamber. This relation underscores the dependency of $\dot{m}$ on combustion chamber conditions.

Adjusting Performance Through Mass Flow Rate Modulation

Tuning the mass flow rate allows for modulation of thrust, which is critical for mission-specific requirements such as stage separation and orbital insertion. This can be achieved via throttleable engines, where variation in $\dot{m}$ is controlled to adjust the thrust output.

Numerically, adjustments to the mass flow rate can be simulated using computational fluid dynamics (CFD) methods. The iterative solution of the Navier-Stokes equations within the nozzle provides insights into how varying $\dot{m}$ impacts the velocity and pressure profiles within the exhaust stream, affecting the overall efficiency and performance metrics of the propulsion system.

Python Code Snippet

Below is a Python code snippet that encompasses the core computational elements involved in mass flow rate calculations, thrust generation, engine performance evaluation, and propellant management.

```python
import numpy as np

def mass_flow_rate(dm, dt):
    '''
    Calculate the mass flow rate.
    :param dm: Differential mass element.
    :param dt: Differential time element.
    :return: Mass flow rate.
    '''
    return dm / dt

def thrust(m_dot, v_e, A, p_e, p_0):
    '''
    Calculate the thrust produced by a rocket engine.
    :param m_dot: Mass flow rate.
    :param v_e: Effective exhaust velocity.
    :param A: Area of the exhaust nozzle.
    :param p_e: Pressure at nozzle exit.
    :param p_0: Ambient pressure.
    :return: Thrust.
    '''
    return m_dot * v_e + A * (p_e - p_0)

def specific_impulse(F, m_dot, g_0=9.81):
```

```python
    '''
    Calculate the specific impulse of a rocket engine.
    :param F: Thrust.
    :param m_dot: Mass flow rate.
    :param g_0: Standard gravity acceleration.
    :return: Specific impulse.
    '''
    return F / (m_dot * g_0)

def nozzle_mass_flow_rate(rho_c, A_c, v_c):
    '''
    Calculate mass flow rate through a nozzle.
    :param rho_c: Propellant density at chamber conditions.
    :param A_c: Chamber cross-sectional area.
    :param v_c: Velocity of the propellant in the chamber.
    :return: Mass flow rate.
    '''
    return rho_c * A_c * v_c

# Example values
dm = 1.0    # kg
dt = 0.1    # s
m_dot = mass_flow_rate(dm, dt)
v_e = 3000    # m/s
A = 1.5    # m^2
p_e = 101325    # Pa
p_0 = 100000    # Pa

F = thrust(m_dot, v_e, A, p_e, p_0)
isp = specific_impulse(F, m_dot)

rho_c = 1.225    # kg/m^3
A_c = 0.5    # m^2
v_c = 100    # m/s
nozzle_flow_rate = nozzle_mass_flow_rate(rho_c, A_c, v_c)

print("Mass Flow Rate:", m_dot)
print("Thrust:", F)
print("Specific Impulse:", isp)
print("Nozzle Mass Flow Rate:", nozzle_flow_rate)
```

This code defines several key functions necessary for understanding and modeling rocket mass flow and performance:

- `mass_flow_rate` calculates the mass flow rate based on differential changes in mass and time.

- `thrust` computes the thrust generated by the rocket engine, incorporating both momentum and pressure thrust elements.

- `specific_impulse` provides a measure of engine efficiency expressed in terms of thrust per unit weight flow rate.

- `nozzle_mass_flow_rate` calculates the mass flow rate as related to nozzle and chamber parameters, emphasizing relations with propellant density and conditions.

The last section of the code offers example calculations using assumed values to demonstrate the functions.

Chapter 5

Specific Impulse (Isp)

Definition and Mathematical Representation

Specific impulse (I_{sp}) serves as a critical parameter in the evaluation of rocket engine efficiency. It is mathematically defined as the impulse per unit weight flow rate of propellant. This efficiency metric is expressed in seconds and describes how effectively a propulsion system can convert propellant mass into thrust, directly correlating to the engine's operational performance within the bounds of its respective propulsion technology.

For a given rocket engine, specific impulse can be calculated as:

$$I_{sp} = \frac{F}{\dot{m} \cdot g_0} \tag{5.1}$$

where: - F represents the total thrust force produced by the engine (in newtons), - $\dot{m}$ is the mass flow rate of the propellant (in kilograms per second), - g_0 is the standard acceleration due to gravity (9.81 m/s^2), providing consistency across various gravitational contexts.

Applications in Rocketry

Specific impulse applies extensively across different types of propulsion systems, notably in measuring efficiency differences between chemical, electric, and hybrid propulsion technologies. This parameter remains fundamental to performance assessment, whether in achieving optimal fuel consumption rates for short-haul missions or ensuring adequate thrust-to-weight ratios for deep space exploration.

Comparative Analysis Across Propulsion Technologies

Distinct propulsion systems demonstrate varying specific impulses:

1 Chemical Propulsion

Chemical rocket engines, harnessing rapid combustion of propellants, achieve modest I_{sp} values ranging typically from 200 to 450 seconds. The performance here primarily depends on the exhaust velocity v_e achieved by the specific propellant chemistry and combustion dynamics, wherein:

$$v_e = I_{sp} \cdot g_0 \qquad (5.2)$$

2 Electric Propulsion

Electric propulsion systems, including ion and Hall effect thrusters, attain considerably higher I_{sp} values, often exceeding 2000 seconds. These higher efficiencies result from the controlled expulsion of ionized particles at exceptionally high velocities. The energy-intensive nature of these systems predicates a different thrust output relative to classical chemical engines, necessitating comprehensive efficiency and power availability assessments.

3 Hybrid Propulsion

Hybrid engines combine chemical and electric components to leverage the high thrust-to-mass ratio of chemical systems with the enhanced I_{sp} characteristics of electric propulsion. Performance thus depends intricately on synchronized operation, often resulting in variable specific impulses contingent on mixed phase applications.

Theoretical Implications

Specific impulse directly correlates to theoretical limitations imposed by thermodynamic cycles and propulsion material restrictions. The Tsiolkovsky rocket equation elucidates these bounds further, demonstrating the ultimate velocity achievable for a given propulsion system:

$$\Delta v = I_{sp} \cdot g_0 \cdot \ln\left(\frac{m_0}{m_f}\right) \qquad (5.3)$$

where m_0 and m_f represent the initial and final total mass of the rocket, including propellant expenditures.

Integration into System Design

Incorporating specific impulse into broader system design involves maximizing engineering heuristics and system constraints. Computing effective exhaust velocities aligns propulsion schematic iterations with desired mission profiles, propelling advancements in contemporary launch technologies. Propellant efficiency categories and cycle configurations dynamically influence the system synthesis process, demanding iterative and analytical engineering methodologies.

Calculation Examples with Python Code

Python algorithms facilitate specific impulse computations for scalable implementation within aerospace engineering tasks. A foundational algorithm enabling practitioners to calculate I_{sp} may appear as follows:

```
def specific_impulse(thrust, mass_flow_rate, g_0=9.81):
    '''
    Calculate the specific impulse of a rocket engine.
    :param thrust: The thrust produced by the rocket engine (N).
    :param mass_flow_rate: The mass flow rate of the propellant
      ↪ (kg/s).
    :param g_0: Standard acceleration due to gravity (9.81 m/s^2).
    :return: Specific impulse (s).
    '''
    return thrust / (mass_flow_rate * g_0)
```

```python
# Example values
thrust = 500000    # N
mass_flow_rate = 250    # kg/s

isp = specific_impulse(thrust, mass_flow_rate)

print("Specific Impulse:", isp)
```

This code encapsulates the principle relationship defining specific impulse, allowing for operation-specific customization dependent on thrust and flow dynamics.

Python Code Snippet

Below is a Python code snippet that encompasses the core computational elements for calculating the specific impulse (I_{sp}) of a rocket engine, as well as related concepts such as exhaust velocity and Δv using the Tsiolkovsky rocket equation.

```python
import numpy as np

def specific_impulse(thrust, mass_flow_rate, g_0=9.81):
    '''
    Calculate the specific impulse of a rocket engine.
    :param thrust: The thrust produced by the rocket engine (N).
    :param mass_flow_rate: The mass flow rate of the propellant
    ↪  (kg/s).
    :param g_0: Standard acceleration due to gravity (9.81 m/s^2).
    :return: Specific impulse (s).
    '''
    return thrust / (mass_flow_rate * g_0)

def exhaust_velocity(Isp, g_0=9.81):
    '''
    Calculate the exhaust velocity.
    :param Isp: Specific impulse (s).
    :param g_0: Standard acceleration due to gravity (9.81 m/s^2).
    :return: Exhaust velocity (m/s).
    '''
    return Isp * g_0

def tsiolkovsky_rocket_equation(Isp, m_initial, m_final, g_0=9.81):
    '''
    Calculate the delta-v using Tsiolkovsky's rocket equation.
    :param Isp: Specific impulse (s).
    :param m_initial: Initial total mass of the rocket including
    ↪  propellant (kg).
    :param m_final: Final total mass of the rocket after burning
    ↪  propellant (kg).
```

```python
    :param g_0: Standard acceleration due to gravity (9.81 m/s^2).
    :return: Delta-v (m/s).
    '''
    return Isp * g_0 * np.log(m_initial / m_final)

# Example values
thrust = 500000       # N
mass_flow_rate = 250  # kg/s
m_initial = 10000     # kg
m_final = 8000        # kg

isp = specific_impulse(thrust, mass_flow_rate)
v_e = exhaust_velocity(isp)
delta_v = tsiolkovsky_rocket_equation(isp, m_initial, m_final)

print("Specific Impulse:", isp, "s")
print("Exhaust Velocity:", v_e, "m/s")
print("Delta-v:", delta_v, "m/s")
```

This code defines several key functions necessary for understanding and calculating specific impulse and its implications in rocketry:

- `specific_impulse` function computes the specific impulse of a rocket engine based on the thrust it produces and the mass flow rate of the propellant.

- `exhaust_velocity` calculates the exhaust velocity from the specific impulse, illustrating how effectively the engine can accelerate exhaust gases.

- `tsiolkovsky_rocket_equation` uses the Tsiolkovsky rocket equation to determine the theoretical Δv capability, given the specific impulse and mass ratio of the rocket.

The final block of code provides examples of computing these elements using typical rocket engine parameters.

Chapter 6

Thrust Equation for Chemical Rockets

Thrust Equation Fundamentals

Chemical rockets function by expelling high-velocity gases through nozzles, generating thrust. The thrust (F) produced by a chemical rocket engine is fundamentally dependent on the mass flow rate of the propellant, the velocity of the exhaust gases, and the pressure differential across the nozzle.

$$F = \dot{m} v_e + (p_e - p_0) A_e \qquad (6.1)$$

where: - $\dot{m}$ represents the mass flow rate of the propellant, - v_e denotes the exhaust velocity, - p_e is the exhaust pressure, - p_0 is the ambient pressure, and - A_e is the exit area of the nozzle.

Exhaust Velocity and Its Determination

The exhaust velocity (v_e) plays a crucial role in the magnitude of the thrust and is determined by the energy conversion processes within the combustion chamber. The relationship between exhaust velocity and other physical parameters is captured by the equation:

$$v_e = \sqrt{\frac{2 \cdot k}{k-1} \cdot \frac{R \cdot T_c}{M_w} \cdot \left[1 - \left(\frac{p_e}{p_c}\right)^{\frac{k-1}{k}}\right]} \qquad (6.2)$$

where: - k is the specific heat ratio, - R is the universal gas constant, - T_c is the combustion chamber temperature, - M_w is the molar mass of the exhaust gases, - p_c is the chamber pressure, and - p_e is the exhaust pressure at the nozzle exit.

These parameters illustrate the comprehensive nature of the processes transforming chemical energy into kinetic energy, governed by isentropic expansion principles.

Chamber Pressure Effects

Chamber pressure (p_c) significantly influences the propulsion system's performance, affecting both exhaust velocity and overall thrust. The chamber pressure is crucial in determining the operational range of pressure differentials that the vehicle can withstand, directly impacting the mass flow rate:

$$\dot{m} = \frac{p_c \cdot A^*}{\sqrt{T_c}} \cdot \left(\frac{k}{R}\right)^{0.5} \cdot \left(\frac{2}{k+1}\right)^{\frac{k+1}{2(k-1)}} \tag{6.3}$$

where A^* is the throat area of the nozzle. This expression is central to quantifying mass flow rate dynamics, thereby directly influencing elaborations in specific impulse and thrust.

Pressure Differential Contributions

The pressure differential term, $(p_e - p_0)A_e$, contributes to thrust generation by considering the forces distributed across the nozzle structure. Elevated chamber pressures typically translate to increased exhaust pressures, modifying the effective impulse delivered by the propulsion system.

Chemical propulsion designs maximize the efficiency of pressure differential effects through optimized nozzle geometries, accounting for variable atmospheric conditions encountered during ascent.

Implications for Propulsion Efficiency

Understanding the interplay between these elements within the thrust equation enables enhancements in propulsion efficiency and engine performance. Chemical propulsion systems capitalize on these dynamics to facilitate optimal mission execution across varying operational envelopes.

Python Code Snippet

Below is a Python code snippet that encapsulates the essential computational components related to chemical rocket thrust equation calculations, including mass flow rate, exhaust velocity, thrust determination, and pressure differential contributions.

```python
import numpy as np

def calculate_thrust(m_dot, v_e, p_e, p_0, A_e):
    '''
    Calculate the thrust produced by a chemical rocket engine.
    :param m_dot: Mass flow rate of the propellant.
    :param v_e: Exhaust velocity.
    :param p_e: Exhaust pressure.
    :param p_0: Ambient pressure.
    :param A_e: Nozzle exit area.
    :return: Thrust value.
    '''
    return m_dot * v_e + (p_e - p_0) * A_e

def determine_exhaust_velocity(k, R, T_c, M_w, p_e, p_c):
    '''
    Determine the exhaust velocity of the rocket.
    :param k: Specific heat ratio.
    :param R: Universal gas constant.
    :param T_c: Combustion chamber temperature.
    :param M_w: Molar mass of exhaust gases.
    :param p_e: Exhaust pressure.
    :param p_c: Chamber pressure.
    :return: Exhaust velocity.
    '''
    return np.sqrt((2 * k / (k - 1)) * (R * T_c / M_w) *
                   (1 - (p_e / p_c)**((k - 1) / k)))

def mass_flow_rate(p_c, A_star, T_c, k, R):
    '''
    Calculate the mass flow rate of the propellant.
    :param p_c: Chamber pressure.
    :param A_star: Throat area of the nozzle.
    :param T_c: Combustion chamber temperature.
    :param k: Specific heat ratio.
    :param R: Universal gas constant.
    :return: Mass flow rate.
    '''
    return (p_c * A_star / np.sqrt(T_c)) * ((k / R)**0.5) * ((2 / (k
        + 1))**((k + 1) / (2 * (k - 1))))

# Define parameters for demonstration
k = 1.4  # Specific heat ratio
R = 8.314  # Universal gas constant
```

```python
T_c = 3500   # Combustion chamber temperature in Kelvin
M_w = 22.7   # Molar mass of exhaust gases, example value
p_e = 0.1    # Exhaust pressure in atm
p_c = 50     # Chamber pressure in atm
A_e = 2.5    # Nozzle exit area in m^2
A_star = 1.0 # Throat area in m^2
p_0 = 0.1    # Ambient pressure in atm

# Calculating exhaust velocity
v_e = determine_exhaust_velocity(k, R, T_c, M_w, p_e, p_c)
print("Exhaust Velocity:", v_e)

# Calculating mass flow rate
m_dot = mass_flow_rate(p_c, A_star, T_c, k, R)
print("Mass Flow Rate:", m_dot)

# Calculating thrust
thrust = calculate_thrust(m_dot, v_e, p_e, p_0, A_e)
print("Thrust Produced:", thrust)
```

This code provides several critical functions for computing chemical rocket thrust:

- `calculate_thrust` determines the thrust generated, considering mass flow rate, exhaust velocity, and pressures.

- `determine_exhaust_velocity` calculates exhaust velocity based on thermodynamic parameters and chamber conditions.

- `mass_flow_rate` estimates the mass flow rate impacted by chamber pressure and throat area.

The accompanying code block demonstrates these calculations using sample data, reflecting the core dynamics of chemical propulsion systems.

Chapter 7

Thrust Equation for Electric Propulsion

Introduction to Electric Propulsion Systems

Electric propulsion alters traditional propulsion methodologies by leveraging electric and magnetic fields to accelerate propellants, notably ions, and achieve thrust. This chapter delves into the specific thrust equations integral to electric propulsion, including analysis pertinent to ion and Hall effect thrusters.

Ion Thruster Thrust Fundamentals

Ion thrusters distinguish themselves by expelling ions to create thrust. The basic thrust equation for an ion thruster is expressed as:

$$F = \dot{m} v_e = I^+ \frac{2qU}{m_i} \tag{7.1}$$

Here: - $\dot{m}$ denotes mass flow rate of ions, - v_e is exhaust velocity, - I^+ represents the ion current, - q is the electron charge, - U is the accelerating potential, and - m_i is the ion mass.

Ion beam neutrality is maintained via a stream of electrons emitted from a neutralizer, balancing charge and ensuring stable thrust.

Exhaust Velocity in Ion Propulsion

The exhaust velocity (v_e) in ion thrusters can be deduced from the energy balance of accelerated ions:

$$v_e = \sqrt{\frac{2qU}{m_i}} \qquad (7.2)$$

This relationship signifies how accelerating potential U, rather than thermal processes, determines ion exhaust speed in electric propulsion systems.

Hall Effect Thruster Dynamics

Hall effect thrusters utilize electric and magnetic fields to confine electrons, which in turn ionize a propellant (commonly Xenon) through electron bombardment, creating ions expelled for thrust:

$$F = \dot{m}v_d + \frac{\Delta p}{A_e} \qquad (7.3)$$

Where: - v_d represents drift velocity of ions under combined electric and magnetic fields.

Mass Flow Rate Calculations

For electric thrusters, mass flow rate ($\dot{m}$) is pivotal in predicting thrust and energy efficiency, defined:

$$\dot{m} = \frac{I^+}{q} \qquad (7.4)$$

This requires precision in ion stream measurements, affecting both velocity and thrust magnitude.

Ionization and Neutralization Efficiency

Efficiency metrics in electric propulsion relate directly to ionization (η_i) and neutralization (η_n) efficiencies, thereby affecting the final thrust performance. These efficiencies characterize the proportion of propellant ions effectively contributing to thrust versus losses due to recombination or inefficient acceleration.

$$\eta = \eta_i \cdot \eta_n \qquad (7.5)$$

Electric propulsion systems target these efficiencies for optimization, ensuring maximal thrust per input energy.

Magnetized Plasma Interactions in Hall Effect Thrusters

In Hall effect thrusters, electron confinement via magnetic fields (B) induces azimuthal E x B drift, facilitating robust ionization kinetics and thrust formation:

$$v_d = \frac{E}{B} \qquad (7.6)$$

This E x B drift is crucial in determining thruster performance characteristics including ion flux stability and directed ion velocity.

Thrust Characteristics and Optimization

Optimization of thrust in electric propulsion pivots around parameter adjustments including ion current (I^+), electromagnetic field calibration, and propellant feed efficiency to achieve desired performance profiles:

$$\texttt{Optimization}(F) = \texttt{maximize}(I^+, U, \eta) \qquad (7.7)$$

Through iterations and testing in electric fields (E) and magnetic (B) field configurations, engineers seek thruster designs balancing thrust magnitude with energy consumption.

Energy and Power Considerations

The electric power (P_e) utilized in electric thrusters is directly related to the kinetic energy of expelled ions:

$$P_e = I^+ U \qquad (7.8)$$

Managing power systems to efficiently supply this electric energy is central to propulsion system design, linking spacecraft energy architecture to operational capacity.

Python Code Snippet

Below is a Python code snippet that encompasses the core computational elements of electric propulsion system calculations, including thrust computation for ion and Hall effect thrusters, exhaust velocity determination, mass flow rate, efficiency, and power considerations.

```python
import numpy as np

def ion_thrust(I_plus, U, q, m_i):
    '''
    Calculate thrust produced by an ion thruster.
    :param I_plus: Ion current (A).
    :param U: Accelerating potential (V).
    :param q: Electron charge (C).
    :param m_i: Ion mass (kg).
    :return: Thrust force (N).
    '''
    return I_plus * np.sqrt(2 * q * U / m_i)

def exhaust_velocity(U, q, m_i):
    '''
    Calculate the exhaust velocity for ion thrusters.
    :param U: Accelerating potential (V).
    :param q: Electron charge (C).
    :param m_i: Ion mass (kg).
    :return: Exhaust velocity (m/s).
    '''
    return np.sqrt(2 * q * U / m_i)

def hall_effect_thrust(m_dot, v_d, delta_p, A_e):
    '''
    Calculate thrust for a Hall effect thruster.
    :param m_dot: Mass flow rate (kg/s).
    :param v_d: Drift velocity (m/s).
    :param delta_p: Pressure difference (N/m^2).
    :param A_e: Exit area (m^2).
    :return: Thrust force (N).
    '''
    return m_dot * v_d + delta_p / A_e

def mass_flow_rate(I_plus, q):
    '''
    Calculate the mass flow rate in an electric thruster.
    :param I_plus: Ion current (A).
    :param q: Electron charge (C).
    :return: Mass flow rate (kg/s).
    '''
    return I_plus / q
```

```python
def efficiency(eta_i, eta_n):
    '''
    Calculate the propulsion system efficiency.
    :param eta_i: Ionization efficiency.
    :param eta_n: Neutralization efficiency.
    :return: Total efficiency.
    '''
    return eta_i * eta_n

def drift_velocity(E, B):
    '''
    Calculate the drift velocity in Hall effect thrusters.
    :param E: Electric field strength (V/m).
    :param B: Magnetic field strength (T).
    :return: Drift velocity (m/s).
    '''
    return E / B

def electric_power(I_plus, U):
    '''
    Calculate the electrical power used by the thruster.
    :param I_plus: Ion current (A).
    :param U: Accelerating potential (V).
    :return: Electrical power (W).
    '''
    return I_plus * U

# Example parameters
I_plus = 2.0    # Ampere
U = 3000.0      # Volt
q = 1.602e-19   # Coulombs
m_i = 2.18e-25  # kg (for Xenon ion)
E = 200.0       # V/m
B = 0.1         # Tesla
m_dot = 5.0e-6  # kg/s
v_d = 16000.0   # m/s
delta_p = 1.0   # N/m^2
A_e = 0.05      # m^2
eta_i = 0.9
eta_n = 0.9

# Calculations
thrust_ion = ion_thrust(I_plus, U, q, m_i)
v_e = exhaust_velocity(U, q, m_i)
thrust_hall = hall_effect_thrust(m_dot, v_d, delta_p, A_e)
mass_flow = mass_flow_rate(I_plus, q)
system_efficiency = efficiency(eta_i, eta_n)
v_drift = drift_velocity(E, B)
power_consumed = electric_power(I_plus, U)

print("Ion Thruster Thrust:", thrust_ion)
print("Exhaust Velocity:", v_e)
print("Hall Effect Thrust:", thrust_hall)
```

```
print("Mass Flow Rate:", mass_flow)
print("System Efficiency:", system_efficiency)
print("Drift Velocity:", v_drift)
print("Electrical Power:", power_consumed)
```

This code defines several key functions necessary for the implementation of electric propulsion systems:

- `ion_thrust` function computes the thrust generated by ion thrusters.

- `exhaust_velocity` calculates the ion exhaust velocity utilizing accelerating potential.

- `hall_effect_thrust` accounts for the thrust generation in Hall effect thrusters.

- `mass_flow_rate` provides mass flow rate calculations important for predicting thrust and energy efficiency.

- `efficiency` determines the propulsion system's total efficiency based on ionization and neutralization metrics.

- `drift_velocity` calculates the azimuthal drift velocity of ions in Hall effect thrusters.

- `electric_power` evaluates the electric power consumption essential for propulsion operation.

By applying these functions with example parameters relevant to ion and Hall effect propulsion, this Python code snippet provides insights into achieving optimized electric propulsion performance.

Chapter 8

Vacuum and Sea Level Thrust

Introduction to Thrust Calculations

Thrust is a fundamental parameter in rocket propulsion. It is defined as the force exerted by a rocket engine, typically measured in newtons (N). Thrust is dynamic and changes with altitude due to variations in atmospheric pressure. This chapter elaborates on the differentiation between vacuum and sea-level thrust, exploring their implications for rocket performance across atmospheric phases.

Sea Level Thrust

Rockets initiate their journey within the Earth's atmosphere, where atmospheric pressure P_a substantially affects thrust generation.

Sea-level thrust (F_{sl}) can be specified as:

$$F_{sl} = \dot{m}v_e + A_e(P_e - P_a)$$

where:

- $\dot{m}$ denotes the mass flow rate in kg/s,
- v_e is the effective exhaust velocity in m/s,
- A_e represents the nozzle exit area in m^2,

- P_a stands for the ambient pressure at sea level in N/m^2,
- P_e represents the pressure at the nozzle exit in N/m^2.

The ambient pressure at sea level (P_a) must be considered when determining the engine's thrust output. Consequently, for sea-level engines, the pressure term $A_e(P_e - P_a)$ acts as a pivotal adjustment factor, emphasizing the interaction between exhaust and atmospheric pressures in producing thrust.

Vacuum Thrust

As a rocket ascends to higher altitudes, the ambient pressure (P_a) approaches zero, effectively resulting in vacuum conditions. Thus, for vacuum thrust (F_v), the expression simplifies to:

$$F_v = \dot{m}v_e + A_e P_e$$

In the absence of atmospheric pressure P_a, the thrust is strictly a function of the effective exhaust velocity and the nozzle exit pressure. Vacuum thrust is inherently more efficient, with the full nozzle expansion term $A_e P_e$ contributing to thrust progress without degradation from ambient pressure opposition.

Nozzle Design and Expansion Ratio

Nozzle design is critical in optimizing engine performance both at sea level and in vacuum conditions. The nozzle expansion ratio (ϵ), defined as:

$$\epsilon = \frac{A_e}{A_t}$$

where A_t is the throat area, becomes a pivotal factor in determining the engine's adaptability across various atmospheric stages. Tailoring the nozzle expansion ratio ideally manages the trade-offs between sea-level and vacuum thrust production.

Thrust Coefficient Analysis

Thrust coefficient (C_f) is a non-dimensional performance metric that correlates with the efficiency and design nuances of rocket engines, given by:

$$C_f = \frac{F}{A_t P_c}$$

where:

- F symbolizes thrust force,
- A_t denotes the throat area, and
- P_c is the chamber pressure.

By determining C_f in both sea level and vacuum conditions, one can discern the influence of atmospheric pressure on engine performance, allowing for adjustments in operational strategies.

Role of Specific Impulse

Specific impulse (I_{sp}) remains a critical evaluation metric for thrust across different phases:

$$I_{sp} = \frac{F}{\dot{m} g_0}$$

where g_0 is the standard gravitational acceleration. Differentiating between $I_{sp,sl}$ (sea level) and $I_{sp,v}$ (vacuum) exposes shifts in propulsion efficiency, guiding design and operational decisions critical to mission success.

Implications for Launch Vehicle Design

Ultimately, mapping the distinctions between vacuum and sea-level thrust informs the architectural selection and operational planning of rocket systems. Considering these thrust variations allows for tailored engine designs capable of transitioning effectively through atmospheric phases, maximizing efficiency and performance alignments across the spectrum of launch vehicle objectives.

Python Code Snippet

Below is a Python code snippet that encompasses the core computational elements for thrust calculations across sea-level and vacuum conditions, nozzle design deployment, and related performance metrics in rocket propulsion.

```python
import numpy as np

def sea_level_thrust(mass_flow_rate, exhaust_velocity, exit_area,
                     exit_pressure, ambient_pressure):
    """
    Calculate the sea-level thrust of a rocket engine.
    :param mass_flow_rate: Mass flow rate (kg/s)
    :param exhaust_velocity: Effective exhaust velocity (m/s)
    :param exit_area: Nozzle exit area (m^2)
    :param exit_pressure: Pressure at the nozzle exit (N/m^2)
    :param ambient_pressure: Ambient pressure at sea level (N/m^2)
    :return: Sea level thrust (N)
    """
    return mass_flow_rate * exhaust_velocity + exit_area * \
        (exit_pressure - ambient_pressure)

def vacuum_thrust(mass_flow_rate, exhaust_velocity, exit_area,
                  exit_pressure):
    """
    Calculate the vacuum thrust of a rocket engine.
    :param mass_flow_rate: Mass flow rate (kg/s)
    :param exhaust_velocity: Effective exhaust velocity (m/s)
    :param exit_area: Nozzle exit area (m^2)
    :param exit_pressure: Pressure at the nozzle exit (N/m^2)
    :return: Vacuum thrust (N)
    """
    return mass_flow_rate * exhaust_velocity + exit_area * \
        exit_pressure

def nozzle_expansion_ratio(exit_area, throat_area):
    """
    Calculate the nozzle expansion ratio.
    :param exit_area: Nozzle exit area (m^2)
    :param throat_area: Throat area (m^2)
    :return: Expansion ratio
    """
    return exit_area / throat_area

def thrust_coefficient(thrust, throat_area, chamber_pressure):
    """
    Calculate the thrust coefficient of a rocket engine.
    :param thrust: Thrust force (N)
    :param throat_area: Throat area (m^2)
    :param chamber_pressure: Chamber pressure (N/m^2)
    :return: Thrust coefficient (non-dimensional)
    """
    return thrust / (throat_area * chamber_pressure)

def specific_impulse(thrust, mass_flow_rate, g0=9.81):
    """
    Calculate the specific impulse of a rocket engine.
    :param thrust: Thrust force (N)
```

```
    :param mass_flow_rate: Mass flow rate (kg/s)
    :param g0: Standard gravitational acceleration (m/s^2)
    :return: Specific impulse (s)
    """
    return thrust / (mass_flow_rate * g0)

# Example parameters
mass_flow_rate = 250      # kg/s
exhaust_velocity = 3000   # m/s
exit_area = 1.5           # m^2
exit_pressure = 10000     # N/m^2
ambient_pressure = 101325 # N/m^2 (sea level)
throat_area = 0.5         # m^2
chamber_pressure = 3000000 # N/m^2

# Calculate sea-level and vacuum thrust
sea_level_thrust_value = sea_level_thrust(mass_flow_rate,
↪  exhaust_velocity, exit_area, exit_pressure, ambient_pressure)
vacuum_thrust_value = vacuum_thrust(mass_flow_rate,
↪  exhaust_velocity, exit_area, exit_pressure)

# Calculate nozzle expansion ratio and thrust coefficient
expansion_ratio = nozzle_expansion_ratio(exit_area, throat_area)
thrust_coeff = thrust_coefficient(sea_level_thrust_value,
↪  throat_area, chamber_pressure)

# Calculate specific impulse
specific_impulse_sea_level =
↪  specific_impulse(sea_level_thrust_value, mass_flow_rate)
specific_impulse_vacuum = specific_impulse(vacuum_thrust_value,
↪  mass_flow_rate)

print("Sea Level Thrust:", sea_level_thrust_value)
print("Vacuum Thrust:", vacuum_thrust_value)
print("Nozzle Expansion Ratio:", expansion_ratio)
print("Thrust Coefficient:", thrust_coeff)
print("Specific Impulse (Sea Level):", specific_impulse_sea_level)
print("Specific Impulse (Vacuum):", specific_impulse_vacuum)
```

This code defines several key functions for estimating the performance of rocket engines across different operational phases:

- `sea_level_thrust` and `vacuum_thrust` functions compute the thrust forces accounting for atmospheric pressure variations.

- `nozzle_expansion_ratio` calculates the ratio essential for adapting nozzles to varied atmospheric pressures.

- `thrust_coefficient` offers a performance measure reflecting efficiency influenced by design parameters.

- `specific_impulse` evaluates engine efficiency under sea-level and vacuum conditions.

The provided code block demonstrates the application of these computational functions using example rocket engine parameters.

Chapter 9

Delta-V Budgeting and Mission Planning

Delta-V and Its Significance in Space Missions

Delta-V (ΔV) is a critical parameter in aerospace engineering, representing the total velocity change required by a spacecraft to accomplish its mission. It encapsulates the cumulative effect of all propulsive maneuvers, determining the feasibility and efficiency of mission trajectories. The mastery of ΔV budgeting involves strategically planning these velocity changes in accordance with mission constraints and objectives.

Derivation and Interpretation of the Delta-V Budget

The ΔV concept stems from the Tsiolkovsky rocket equation, expressed as:

$$\Delta V = v_e \ln\left(\frac{m_0}{m_f}\right) \tag{9.1}$$

where v_e is the effective exhaust velocity, m_0 represents the initial mass of the rocket, and m_f is the final mass after fuel consumption. This formula delineates the fundamental relationship between fuel usage and velocity increment, forming the bedrock of mission planning.

Allocation of Delta-V in Mission Phases

Strategically allocating ΔV across distinct mission phases is pivotal. Each phase, including launch, transfer orbits, and insertion burns, necessitates precise velocity augmentations. The allocation must balance fuel reserves against mission objectives, often necessitating trade-offs. High-fidelity mission designs require detailed computation to optimize fuel partitioning across these stages.

Delta-V Requirements for Common Orbital Transfers

Numerous orbital transfers impose specific ΔV demands. The Hohmann transfer, a prominent method for transitioning between coplanar orbits, requires a calculable ΔV:

$$\Delta V_1 = \sqrt{\frac{GM}{r_1}} \left(\sqrt{\frac{2r_2}{r_1 + r_2}} - 1 \right) \tag{9.2}$$

$$\Delta V_2 = \sqrt{\frac{GM}{r_2}} \left(1 - \sqrt{\frac{2r_1}{r_1 + r_2}} \right) \tag{9.3}$$

where r_1 and r_2 are the initial and final orbital radii, and G and M denote the gravitational constant and Earth's mass, respectively. These calculations enable precise ΔV distribution necessary for successful orbital transition.

Efficient Delta-V Utilization in Interplanetary Missions

Interplanetary missions demand meticulous ΔV budgeting due to the expansive distances and gravitational interplay. Gravity assist maneuvers, exploiting planetary motion to impart additional energy, exemplify efficient ΔV use, minimizing propellant expenditure while maximizing trajectory gains. Moreover, the low-thrust propulsion options, such as ion thrusters, integrate into mission designs to achieve desired velocity profiles with enhanced efficiency.

Mission Planning Considerations and Constraints

Mission planning necessitates accounting for a myriad of constraints, including payload mass, spacecraft design limitations, and target celestial body characteristics. Incorporating these variables ensures comprehensive ΔV budgeting that precludes mission discrepancies. Engine performance, burn duration, and orbital alignment further influence planning, necessitating iterative simulations to fine-tune ΔV allocations.

Numerical Methods for Delta-V Optimization

Numerical optimization techniques, such as Lambert's algorithm and direct optimization methods, provide systematic frameworks for solving ΔV allocation problems. These approaches enable the determination of optimal trajectories subject to predefined constraints. Implementations in software, such as the 'numpy' and 'scipy.optimize' libraries, facilitate computational efficiencies and enhance mission planning accuracy through robust simulation.

In these contexts, ΔV serves as a definitive mission parameter, driving the synthesis of spacecraft design and mission architecture with the overarching goal of maximizing velocity efficiency and mission success.

Python Code Snippet

Below is a Python code snippet that encompasses the key computational elements related to delta-V budgeting and mission planning, including the calculation of delta-V requirements for various mission phases and optimizing its usage through numerical methods.

```
import numpy as np
import scipy.optimize as opt

def tsiolkovsky_rocket_eq(v_e, m_0, m_f):
    '''
    Calculate delta-V using the Tsiolkovsky rocket equation.
    :param v_e: Effective exhaust velocity.
    :param m_0: Initial mass of the rocket.
```

```python
    :param m_f: Final mass after fuel consumption.
    :return: Delta-V.
    '''
    return v_e * np.log(m_0 / m_f)

def delta_v_allocation(segments, total_delta_v):
    '''
    Distribute total delta-V across mission phases.
    :param segments: List of mission segments.
    :param total_delta_v: Total delta-V available.
    :return: Delta-V allocation per segment.
    '''
    allocation = [s * total_delta_v / sum(segments) for s in
        segments]
    return allocation

def hohmann_transfer_delta_v(r1, r2, GM):
    '''
    Calculate delta-V for Hohmann transfer.
    :param r1: Initial orbital radius.
    :param r2: Final orbital radius.
    :param GM: Gravitational parameter.
    :return: Delta-V1, Delta-V2.
    '''
    delta_v1 = np.sqrt(GM / r1) * (np.sqrt(2 * r2 / (r1 + r2)) - 1)
    delta_v2 = np.sqrt(GM / r2) * (1 - np.sqrt(2 * r1 / (r1 + r2)))
    return delta_v1, delta_v2

def optimize_trajectory(objective_function, bounds, constraints):
    '''
    Optimize trajectory using a numerical method.
    :param objective_function: The function to minimize.
    :param bounds: Bounds for the variables.
    :param constraints: Constraints for the optimization.
    :return: Optimized parameters.
    '''
    result = opt.minimize(objective_function, bounds=bounds,
        constraints=constraints)
    return result.x if result.success else None

# Example function for optimization (e.g., minimal delta-V
    consumption)
def objective_function_example(params):
    # Placeholder example function; real one would calculate some
        mission cost metric
    return sum(params ** 2)

# Example usage
v_e = 3000      # m/s
m_0 = 5000      # kg
m_f = 3000      # kg
total_delta_v = tsiolkovsky_rocket_eq(v_e, m_0, m_f)
print("Total Delta-V:", total_delta_v)
```

```
segments = [1, 0.5, 1.5]  # Example allocation segments
alloc = delta_v_allocation(segments, total_delta_v)
print("Delta-V Allocation:", alloc)

r1, r2, GM = 7000, 42000, 398600  # Example Hohmann transfer
    parameters
delta_v1, delta_v2 = hohmann_transfer_delta_v(r1, r2, GM)
print("Hohmann Transfer Delta-V1:", delta_v1, "Delta-V2:", delta_v2)

optimized_params = optimize_trajectory(objective_function_example,
    bounds=[(0, 1)] * 3, constraints=[])
print("Optimized Parameters:", optimized_params)
```

This code defines several key functions necessary for delta-V calculations and mission planning:

- `tsiolkosky_rocket_eq` function computes delta-V using the Tsiolkovsky rocket equation.

- `delta_v_allocation` distributes total delta-V across mission segments based on segment priorities.

- `hohmann_transfer_delta_v` calculates the delta-V required for a Hohmann transfer orbit.

- `optimize_trajectory` employs numerical optimization to improve mission delta-V usage or other parameters.

- `objective_function_example` is a placeholder for an example objective function used in optimization.

The final block of code provides examples of calculating delta-V for various mission scenarios using simplified parameters and demonstrates a mock task of trajectory optimization.

Chapter 10

Orbital Mechanics Fundamentals

Kepler's Laws of Planetary Motion

Kepler's Laws form the foundation of classical orbital mechanics. The laws describe how celestial bodies orbit in elliptical paths due to gravitational forces, providing essential insights into the trajectories of artificial satellites and spacecraft.

1 First Law: The Law of Ellipses

Kepler's First Law states that the path of a planet around the Sun is an ellipse, with the Sun at one of the two foci. Mathematically, the equation of an ellipse in polar coordinates is given by:

$$r(\theta) = \frac{a(1-e^2)}{1+e\cos\theta} \qquad (10.1)$$

where r is the radius vector, a represents the semi-major axis, e is the eccentricity of the ellipse, and θ is the true anomaly.

2 Second Law: The Law of Equal Areas

Kepler's Second Law, or the Law of Equal Areas, posits that a line segment joining a planet and the Sun sweeps out equal areas in equal intervals of time. This can be expressed as:

$$\frac{dA}{dt} = \frac{1}{2}r^2\frac{d\theta}{dt} = \text{constant} \qquad (10.2)$$

where A is the area swept, r is the distance to the Sun, and $\frac{d\theta}{dt}$ is the angular speed.

3   Third Law: The Harmonic Law

Kepler's Third Law establishes a relationship between the time period T of an orbit and its semi-major axis a:

$$T^2 \propto a^3 \qquad (10.3)$$

For a two-body system under the gravitational influence of a central body with mass M, this becomes:

$$T^2 = \frac{4\pi^2}{GM}a^3 \qquad (10.4)$$

where G denotes the gravitational constant.

Newton's Law of Universal Gravitation

The gravitational attraction between two masses is described by Newton's Law of Universal Gravitation, with the force F between two bodies of mass m_1 and m_2 separated by distance r given by:

$$F = \frac{Gm_1m_2}{r^2} \qquad (10.5)$$

This fundamental principle underlies the dynamics of celestial bodies and informs the equations of motion for orbiting spacecraft.

Equation of Motion for Orbits

Deriving the motion of a satellite in a gravitational field begins with Newton's Second Law in the context of gravitational force:

$$m\frac{d^2\mathbf{r}}{dt^2} = -\frac{GMm}{r^2}\hat{\mathbf{r}} \qquad (10.6)$$

Solving this differential equation yields the orbital trajectory, often expressed in terms of specific energy and angular momentum vectors.

Vis-Viva Equation

The Vis-Viva Equation relates the velocity v of a spacecraft at any point in its orbit to its radial distance r from the central body and the size of the orbit:

$$v^2 = GM \left(\frac{2}{r} - \frac{1}{a} \right) \tag{10.7}$$

This equation is instrumental in calculating the velocity of a spacecraft at various orbital points.

Specific Orbital Energy

For an object in orbit, the specific orbital energy ϵ is a constant and given as:

$$\epsilon = \frac{v^2}{2} - \frac{GM}{r} \tag{10.8}$$

Specific energy is pivotal in determining the bound or unbound nature of an orbit—negative values correspond to elliptical orbits, whereas zero and positive energy indicate parabolic and hyperbolic trajectories, respectively.

Orbital Elements Determination

The description of an orbit in three-dimensional space is facilitated by six classical orbital elements, comprising semi-major axis a, eccentricity e, inclination i, right ascension of the ascending node Ω, argument of periapsis ω, and true anomaly θ. These elements can be inferred through observations and calculated via transformations across various reference frames governed by the equations of motion.

Python Code Snippet

Below is a Python code snippet that encompasses the core computational elements of orbital mechanics fundamentals, illustrating calculations for Kepler's laws, motion equations, and orbital elements.

```python
import numpy as np

def keplers_first_law(semi_major_axis, eccentricity, true_anomaly):
    '''
    Calculate the radius vector according to Kepler's First Law.
    :param semi_major_axis: Semi-major axis of the orbit.
    :param eccentricity: Eccentricity of the orbit.
    :param true_anomaly: Current angle of the orbiting body.
    :return: Radius vector r.
    '''
    return semi_major_axis * (1 - eccentricity ** 2) / (1 +
    ↪    eccentricity * np.cos(true_anomaly))

def keplers_second_law(constant_area_rate, radius, angular_speed):
    '''
    Calculate the constant rate of area swept according to Kepler's
    ↪    Second Law.
    :param constant_area_rate: Formula component constant for equal
    ↪    area.
    :param radius: Distance from the central body.
    :param angular_speed: Angular speed of the body.
    :return: Rate of area swept.
    '''
    return constant_area_rate / (radius ** 2 * angular_speed)

def keplers_third_law(semi_major_axis, central_body_mass):
    '''
    Calculate the period of orbit using Kepler's Third Law.
    :param semi_major_axis: Semi-major axis of the orbit.
    :param central_body_mass: Mass of the central body.
    :return: Orbital period.
    '''
    G = 6.67430e-11  # gravitational constant
    return 2 * np.pi * np.sqrt((semi_major_axis ** 3) / (G *
    ↪    central_body_mass))

def universal_gravitation(m1, m2, distance):
    '''
    Calculate gravitational force between two masses.
    :param m1: Mass of the first body.
    :param m2: Mass of the second body.
    :param distance: Distance between the two masses.
    :return: Gravitational force.
    '''
    G = 6.67430e-11  # gravitational constant
    return G * m1 * m2 / distance ** 2

def orbital_velocity(central_body_mass, distance, semi_major_axis):
    '''
    Calculate velocity at a point in orbit using the vis-viva
    ↪    equation.
    :param central_body_mass: Mass of the central body.
```

```python
    :param distance: Current distance of the orbiting body.
    :param semi_major_axis: Semi-major axis of the orbit.
    :return: Orbital velocity.
    '''
    G = 6.67430e-11  # gravitational constant
    return np.sqrt(G * central_body_mass * (2 / distance - 1 /
    ↪ semi_major_axis))

def specific_orbital_energy(velocity, central_body_mass, distance):
    '''
    Calculate the specific orbital energy of a body.
    :param velocity: Velocity of the orbiting body.
    :param central_body_mass: Mass of the central body.
    :param distance: Distance from the central body.
    :return: Specific orbital energy.
    '''
    G = 6.67430e-11  # gravitational constant
    return (velocity ** 2) / 2 - (G * central_body_mass) / distance

# Example calculations with hypothetical data
semi_major_axis = 7000e3  # meters
eccentricity = 0.001
true_anomaly = np.deg2rad(30)  # radians
central_body_mass = 5.972e24  # kg (Earth)
m1, m2 = 1000, 5.972e24  # kg
distance = 7000e3  # meters

radius_vector = keplers_first_law(semi_major_axis, eccentricity,
↪ true_anomaly)
orbital_period = keplers_third_law(semi_major_axis,
↪ central_body_mass)
gravity_force = universal_gravitation(m1, m2, distance)
orbital_velocity_value = orbital_velocity(central_body_mass,
↪ distance, semi_major_axis)
orbit_energy = specific_orbital_energy(orbital_velocity_value,
↪ central_body_mass, distance)

print("Radius Vector:", radius_vector)
print("Orbital Period:", orbital_period)
print("Gravitational Force:", gravity_force)
print("Orbital Velocity:", orbital_velocity_value)
print("Specific Orbital Energy:", orbit_energy)
```

This code defines several key functions necessary for the computational aspects of orbital mechanics:

- `keplers_first_law` calculates the radius vector according to the elliptical orbit defined by Kepler's First Law.
- `keplers_second_law` determines the constant rate at which orbital area is swept over time, although typically demonstrated conceptually rather than computationally.

- `keplers_third_law` computes the orbital period utilizing the relationship between period and semi-major axis.

- `universal_gravitation` provides the gravitational force between two masses.

- `orbital_velocity` uses the vis-viva equation to compute the orbital velocity at any point.

- `specific_orbital_energy` calculates the specific energy of an orbit, indicating the dynamic status of the orbit.

The example section demonstrates calculations using hypothetical orbital data.

Chapter 11

Hohmann Transfer Orbit

Introduction to Hohmann Transfer Orbits

The Hohmann transfer orbit is a fundamental concept in astrodynamics, facilitating efficient spacecraft transfer between two coplanar circular orbits. This maneuver is characterized by its minimal energy requirement when addressing orbit changes. The governing principles of Hohmann transfer rely heavily on the two-impulse transfer strategy, optimizing the change in velocity, or ΔV, necessary for such orbital adjustments.

Mathematical Framework

The derivation of the ΔV for a Hohmann transfer orbit involves crucial orbital mechanics principles, relying on gravitational effects and velocity alterations. Consider two circular orbits with radii r_1 and r_2, the initial and final orbits, respectively. The transfer orbit in this scenario will be an ellipse with a semi-major axis $a = \frac{r_1 + r_2}{2}$.

1 $\Delta V Calculations$

For the Hohmann transfer, the required change in velocity encompasses two instantaneous impulses: one to transition from the initial orbit to the transfer orbit and another to achieve the final orbit. The velocity in the initial orbit v_1 is calculated as:

$$v_1 = \sqrt{\frac{GM}{r_1}}$$

where G is the universal gravitational constant and M is the mass of the central body.

The velocity at the periapsis of the transfer ellipse v_p is given by:

$$v_p = \sqrt{2\frac{GM}{r_1} - \frac{GM}{a}}$$

Thus, the initial ΔV_1 required at periapsis to move from the initial circular orbit to the transfer orbit is:

$$\Delta V_1 = v_p - v_1 = \sqrt{2\frac{GM}{r_1} - \frac{GM}{a}} - \sqrt{\frac{GM}{r_1}}$$

The velocity in the final orbit v_2 is:

$$v_2 = \sqrt{\frac{GM}{r_2}}$$

The velocity at the apoapsis of the transfer orbit v_a is:

$$v_a = \sqrt{2\frac{GM}{r_2} - \frac{GM}{a}}$$

The subsequent ΔV_2 required at apoapsis to transition from the transfer ellipse to the final circular orbit at radius r_2 is:

$$\Delta V_2 = v_2 - v_a = \sqrt{\frac{GM}{r_2}} - \sqrt{2\frac{GM}{r_2} - \frac{GM}{a}}$$

Therefore, the total ΔV for the Hohmann transfer is:

$$\Delta V = \Delta V_1 + \Delta V_2$$

2 Efficiency and Constraints

The simplicity of the Hohmann transfer offers a means of inter-orbital transit with minimal propellant expenditure, but it presumes the coplanarity of the initial and final orbits, as well as significant idealization of the physical environment, such as neglect of non-impulsive thrust periods. The method is largely effective for coplanar orbit transfers within a singular gravitational influence, suitable for space logistics, including geostationary satellite insertion and interplanetary trajectories.

Application in Interplanetary Missions

The practicality of the Hohmann trajectory extends to interplanetary travel, where it forms the backbone for mission planning to outer planets. When applying Hohmann transfer principles to spacecraft intended for celestial bodies beyond Earth's sphere, attention shifts to launch windows and available propellant mass.

Interplanetary Hohmann maneuvers require aligning Earth's orbit with the desired destination point, ensuring the planetary alignment periodically matches the transfer opportunity. In this context, planetary ΔV considerations must account for additional factors, such as gravitational assist and non-Keplerian dynamics.

The alignment and timing are critical in determining the feasibility of a mission, the primary limitation being the time sensitivity embedded within launch window patterns dictated by the relative positions of planetary bodies.

Python Code Snippet

Below is a Python code snippet that encompasses the core computational elements for calculating important variables in a Hohmann transfer orbit, including ΔV calculation, velocity computations for various points in orbits, and illustrating these concepts with a practical example.

```
import numpy as np

def hohmann_transfer_dv(r1, r2, GM):
    '''
    Calculate the total V for a Hohmann transfer orbit.
    :param r1: Radius of the initial circular orbit.
    :param r2: Radius of the final circular orbit.
    :param GM: Standard gravitational parameter (G * M).
    :return: Total V required for the Hohmann transfer.
    '''
    a = (r1 + r2) / 2 # semi-major axis of transfer orbit

    # Velocity in initial orbit
    v1 = np.sqrt(GM / r1)
```

```python
    # Velocity at periapsis of transfer orbit
    vp = np.sqrt(2 * GM / r1 - GM / a)

    # Initial V required at periapsis
    delta_v1 = vp - v1

    # Velocity in final orbit
    v2 = np.sqrt(GM / r2)

    # Velocity at apoapsis of transfer orbit
    va = np.sqrt(2 * GM / r2 - GM / a)

    # Final V at apoapsis
    delta_v2 = v2 - va

    # Total V for Hohmann transfer
    total_delta_v = delta_v1 + delta_v2

    return total_delta_v

# Example values for practical implementation
r1 = 6700e3    # Radius of low Earth orbit in meters
r2 = 42164e3   # Radius of geostationary orbit in meters
GM = 3.986004418e14   # Gravitational parameter for Earth in m^3/s^2

# Calculating the V
total_delta_v = hohmann_transfer_dv(r1, r2, GM)

print(f"Total V for Hohmann transfer: {total_delta_v:.2f} m/s")
```

This code defines several key functions necessary for the calculation of ΔV in a Hohmann transfer orbit:

- `hohmann_transfer_dv` function computes the total ΔV necessary for transferring from one circular orbit to another using a Hohmann transfer orbit.

- The example values illustrate how to apply this function for a scenario transferring between a low Earth orbit and a geostationary orbit.

The final block of code demonstrates the usage of these calculations with sample data, providing a clear value for the total ΔV required in such a transfer, which is crucial for understanding propulsion needs in orbital mechanics.

Chapter 12

Bi-Elliptic Transfer

Conceptualization of Bi-Elliptic Transfer

The bi-elliptic transfer presents an alternative orbital maneuver strategy that can potentially lower the delta-V requirements over the conventional Hohmann transfer for large orbit ratio transfers. This technique consists of three distinct impulse maneuvers, spanning two intermediate elliptical transfer orbits. The specificity of its application lies in the precise optimization of ΔV for transfers where the ratio of the final to initial orbit radii is significantly high.

Mathematical Derivation

The derivation of the bi-elliptic transfer involves examining each impulse separately alongside the orbit determination from one conceptual point to another.

1 Transfer Architecture

1. **First Impulse (ΔV_1)**: Initiates transfer from the initial circular orbit of radius r_1 to an elliptical orbit with an apoapsis at a distant point r_b.

2. **Second Impulse (ΔV_2)**: Occurs at the apoapsis r_b to move to the second elliptical transfer orbit, altering the periapsis to the target orbit radius r_2.

3. **Third Impulse (ΔV_3)**: Finalizes the transfer at periapsis r_2, circularizing the new orbit.

The specific objective of bi-elliptic transfer is a function of minimizing total propellant consumption in scenarios where direct transfers yield inefficient solutions.

2 Velocity and Delta-V Calculations

In a bi-elliptic transfer, velocity calculations for each transfer maneuver are crucial. Define the semi-major axis a_1, a_2, and a_3 for the initial orbit, first transfer orbit, and second transfer orbit, respectively. The velocity calculations are indispensable at each prescribed orbital point. Given the gravitational parameter GM, velocities can be determined as follows:

$$v_1 = \sqrt{\frac{GM}{r_1}}$$

For the velocity at apoapsis on the first transfer ellipse:

$$v_{a1} = \sqrt{2\frac{GM}{r_1} - \frac{GM}{a_1}}, \quad \text{where} \quad a_1 = \frac{r_1 + r_b}{2}$$

Subsequent ΔV_1 is:

$$\Delta V_1 = v_{a1} - v_1$$

At the transition to the second elliptical orbit at r_b, velocity calculations include:

$$v_b = \sqrt{2\frac{GM}{r_b} - \frac{GM}{a_1}}$$

$$v_{a2} = \sqrt{2\frac{GM}{r_b} - \frac{GM}{a_2}}, \quad \text{where} \quad a_2 = \frac{r_b + r_2}{2}$$

Thus, ΔV_2 is:

$$\Delta V_2 = v_{a2} - v_b$$

Finally, the circularization velocity at the second transfer orbit's periapsis:

$$v_2 = \sqrt{\frac{GM}{r_2}}$$

$$v_{p2} = \sqrt{2\frac{GM}{r_2} - \frac{GM}{a_2}}$$

ΔV_3 is defined as:

$$\Delta V_3 = v_2 - v_{p2}$$

The total ΔV becomes the aggregate of all three impulses:

$$\Delta V_{\text{total}} = \Delta V_1 + \Delta V_2 + \Delta V_3$$

Assessment of Efficiency in Transfers

Employing bi-elliptic transfer effectively reduces energy consumption, making it the method of choice for missions engaging in far-reaching orbital transfers, specifically when the radius ratio $\frac{r_2}{r_1} > 11.94$. Optimization in this realm necessitates accurate simulation and calculation due to non-linear dynamics and planetary influences.

Comparative Application in Interplanetary Missions

The utility of bi-elliptic transfer primarily hinges on its applicability to craft interplanetary trajectories where direct energy exchanges are suboptimal. Given the known gravitational fields exerted by celestial bodies, predicting precise ΔV savings might necessitate computational modeling, embracing nuances of celestial mechanics beyond planar analytical structures.

Python Code Snippet

Below is a Python code snippet that encompasses the core computational elements of performing a bi-elliptic transfer, including the calculation of orbital velocities and delta-V values for each impulse required during the transfer.

```
import math

def bi_elliptic_transfer(r1, rb, r2, GM):
    '''
    Calculate the delta-V for a bi-elliptic transfer.
    :param r1: Initial orbit radius.
    :param rb: Apoapsis of the first transfer orbit.
    :param r2: Final orbit radius.
    :param GM: Gravitational parameter (constant for a particular
    ↪ celestial body).
    :return: Total delta-V for the transfer.
    '''
```

```python
    # Initial circular orbit velocity
    v1 = math.sqrt(GM / r1)

    # First impulse - from initial orbit to first ellipse
    a1 = (r1 + rb) / 2
    va1 = math.sqrt(2 * GM / r1 - GM / a1)
    delta_v1 = va1 - v1

    # Second impulse - at apoapsis of first ellipse
    vb = math.sqrt(2 * GM / rb - GM / a1)
    a2 = (rb + r2) / 2
    va2 = math.sqrt(2 * GM / rb - GM / a2)
    delta_v2 = va2 - vb

    # Third impulse - at final orbit
    vp2 = math.sqrt(2 * GM / r2 - GM / a2)
    v2 = math.sqrt(GM / r2)
    delta_v3 = v2 - vp2

    # Total delta-V for the bi-elliptic transfer
    delta_v_total = delta_v1 + delta_v2 + delta_v3

    return delta_v_total

# Example usage
GM_earth = 3.986e14  # Gravitational parameter for Earth in m^3/s^2
r1 = 7000e3  # Initial orbit radius in meters (low Earth orbit approx.)
rb = 50000e3  # Example apoapsis distance in meters
r2 = 15000e3  # Final orbit radius in meters

delta_v = bi_elliptic_transfer(r1, rb, r2, GM_earth)
print("Total Delta-V for Bi-Elliptic Transfer:", delta_v, "m/s")
```

This code defines key functions and their utilization involving:

- `bi_elliptic_transfer` function computes the total delta-V required for completing a bi-elliptic transfer, given the radii of initial, apoapsis, and final orbits, along with the gravitational parameter of the celestial body.

- The function makes separate calculations for each of the three impulses: initiating transfer from the initial circular orbit to the first elliptical orbit, transitioning at the apoapsis, and circularizing at the final orbit.

- Example values for a transfer involving distances typical for low Earth orbit and beyond are provided for demonstration purposes.

Chapter 13

Patched Conic Approximations

Overview of Patched Conic Approximation

The patched conic approximation is an analytical approach to simplifying trajectory calculations in multi-body celestial environments. This technique divides complex orbit propagation into distinct regions, allowing for calculations under the influence of a single dominant gravitational body at any given time. It leverages the superposition of conic sections to provide a piecewise analytical solution, applicable to missions across multiple celestial orbits.

Framework of Patch Points

Within the patched conic approximation, transition points between different gravitational dominions are referred to as patch points. These are critical junctures where the trajectory transitions between two dominant gravitational bodies. Each region is characterized by a set of elliptical, hyperbolic, or parabolic trajectories, all calculable using classical two-body dynamics.

Consider a spacecraft moving from Earth to another planet. The trajectory is divided into three primary segments: Earth-centered orbit, heliocentric orbit, and destination planet-centered orbit. The continuity ensures that the transition velocities are matched at each patch point.

1 Mathematical Formulation

In the analysis, the gravitational reach of each celestial body defines the edges of the conic sectors. The dominating gravitational body in each sector fulfills:

$$F_{\text{gravity}} = \frac{GM_{\text{body}}m}{r^2}$$

where G denotes the universal gravitational constant, M_{body} the mass of the celestial body, m the mass of the spacecraft, and r the radial distance from the body's center.

2 Conic Section Determination

For a two-body problem, the equation of the conic section is given as:

$$r = \frac{p}{1 + e\cos\nu}$$

where p is the semi-latus rectum, e is the eccentricity, and ν is the true anomaly. The nature of the conic section (elliptical, parabolic, or hyperbolic) dictates trajectory characteristics within each region.

Transition Between Conics

1 Earth Departure to Heliocentric Orbit

The transition from Earth orbit to heliocentric orbit can be described via a hyperbolic excess velocity, v_∞:

$$v_\infty = \sqrt{\left(\frac{2GM_{\text{Earth}}}{r_{\text{launch}}} - v_{\text{launch}}^2\right)}$$

where r_{launch} represents the radial distance at launch, and v_{launch} is the launch velocity.

The energy at patch points is preserved using the vis-viva equation:

$$v^2 = GM\left(\frac{2}{r} - \frac{1}{a}\right)$$

where a is the semi-major axis of the trajectory segment.

2 Heliocentric to Target Body Orbit

Upon entry into the destination's gravitational field, the heliocentric velocity relates to the celestial body's velocity through:

$$v_{\text{arrival}} = v_\infty + v_{\text{target}}$$

where v_{target} is the velocity of the target planet. The matching condition at the second patch point is satisfied:

$$\Delta v = v_{\text{arriving}} - v_{\text{target}}$$

Application of Patched Conics

1 Analytical Trajectory Simplification

This conic patching strategy excels in scenarios where analytical simplicity is favored over numerical complexity. By reducing a multi-body problem into a succession of solvable two-body problems, it importantly bridges the computational gap in interplanetary travel.

The practical implication of patched conic approximations includes mission planning for gravity assist maneuvers, where spacecraft leverage celestial bodies to gain additional velocity. The piecewise linearity facilitates analytically tractable equation sets for each segment:

$$v_{\text{gain}} = 2v_{\text{planet}} \cos \delta$$

where δ is the deflection angle relative to the ingress trajectory.

2 Planetary Flyby Calculations

Astrodynamics leveraging patched conics utilize the iteratively calculated flyby velocities to map gravitational assists, crucial for extended mission profiles. By orchestrating a sequence of transition orbits, a spacecraft can incrementally build velocity through calculated gravitational encounters.

Python Code Snippet

Below is a Python code snippet that encompasses the implementation of the patched conic approximation method for trajectory calculations in celestial mechanics, focusing on the transition points and gravitational interactions mentioned throughout the chapter.

```python
import numpy as np

def gravitational_force(G, M, m, r):
    '''
    Calculate the gravitational force.
    :param G: Universal gravitational constant.
    :param M: Mass of the celestial body.
    :param m: Mass of the spacecraft.
    :param r: Radial distance from the celestial body's center.
    :return: Gravitational force.
    '''
    return (G * M * m) / r**2

def conic_section_params(p, e, nu):
    '''
    Determine the conic section parameters.
    :param p: Semi-latus rectum.
    :param e: Eccentricity.
    :param nu: True anomaly.
    :return: Radial distance in the conic section.
    '''
    return p / (1 + e * np.cos(nu))

def hyperbolic_excess_velocity(G, M, r_launch, v_launch):
    '''
    Calculate the hyperbolic excess velocity.
    :param G: Universal gravitational constant.
    :param M: Mass of Earth.
    :param r_launch: Radial distance at launch.
    :param v_launch: Launch velocity.
    :return: Hyperbolic excess velocity.
    '''
    return np.sqrt(2 * G * M / r_launch - v_launch**2)

def vis_viva_equation(G, M, r, a):
    '''
    Apply the vis-viva equation to calculate velocity.
    :param G: Universal gravitational constant.
    :param M: Mass of the celestial body.
    :param r: Radial distance.
    :param a: Semi-major axis.
    :return: Velocity.
    '''
    return np.sqrt(G * M * (2 / r - 1 / a))
```

```python
def velocity_matching(v_arrival, v_target):
    '''
    Calculate velocity matching at the target patch point.
    :param v_arrival: Velocity at arrival.
    :param v_target: Velocity of the target body.
    :return: Delta velocity.
    '''
    return v_arrival - v_target

# Constants
G = 6.67430e-11  # m^3 kg^-1 s^-2
M_earth = 5.972e24  # kg
r_earth = 6371000  # m

# Example parameters for Earth to Mars mission
p = 15000  # semi-latus rectum, m
e = 0.1  # eccentricity
nu = np.radians(30)  # true anomaly in radians
r_launch = 6671000  # m
v_launch = 8000  # m/s

# Calculations
F_gravity = gravitational_force(G, M_earth, 1000, r_launch)
r_conic = conic_section_params(p, e, nu)
v_infinity = hyperbolic_excess_velocity(G, M_earth, r_launch,
↪    v_launch)
v_velocity = vis_viva_equation(G, M_earth, r_launch, 15000)

print("Gravitational Force:", F_gravity, "N")
print("Radial Distance for Conic Section:", r_conic, "m")
print("Hyperbolic Excess Velocity:", v_infinity, "m/s")
print("Velocity from Vis-Viva Equation:", v_velocity, "m/s")
```

This code defines several key functions necessary for the implementation of the patched conic approximation:

- gravitational_force calculates the gravitational force acting on a spacecraft by a celestial body.

- conic_section_params determines the properties of the conic section based on given parameters.

- hyperbolic_excess_velocity computes the velocity required for the spacecraft to leave a celestial body's gravitational influence.

- vis_viva_equation applies the vis-viva equation to find velocity at a point in orbit.

- `velocity_matching` calculates the necessary delta velocity for achieving the desired orbital transition.

The final block of code provides outputs for gravitational force, radial distance in a conic section, hyperbolic excess velocity, and velocity from the vis-viva equation using example parameters.

Chapter 14

Thrust-to-Weight Ratio

Thrust-to-Weight Ratio Fundamentals

The thrust-to-weight ratio (TWR) is a crucial parameter in rocket engineering, signifying the capability of a rocket's engine thrust to overcome gravitational forces acting upon it. It is defined by the ratio $\text{TWR} = \frac{T}{W}$, where T represents the thrust produced by the rocket engines and W is the weight of the rocket. This dimensionless quantity influences the design and performance evaluation across various phases of flight.

Static Thrust-to-Weight Ratio

In static conditions at lift-off, the thrust-to-weight ratio must exceed unity for the rocket to ascend. Assuming T is the total thrust produced by all engines and $W = m_{\text{rocket}} \cdot g_0$ (where m_{rocket} is the total mass of the rocket and g_0 is the standard gravitational acceleration), the static thrust-to-weight ratio is:

$$\text{TWR}_{\text{static}} = \frac{T}{m_{\text{rocket}} \cdot g_0}$$

The value of $\text{TWR}_{\text{static}}$ is pivotal during the initial phase to ensure that the vehicle can overcome gravitational pull while being able to control acceleration and structural loads.

Thrust-to-Weight Ratio Dynamics During Ascent

As the rocket ascends, both thrust (T) and weight (W) vary due to factors such as atmospheric pressure changes affecting engine performance and the continuous expulsion of propellant mass. The changing thrust due to variations in atmospheric pressure is expressed as:

$$T = T_{\text{vacuum}} - A_e \cdot (P_{\text{vacuum}} - P_a)$$

where T_{vacuum} is the engine thrust in vacuum conditions, A_e is the engine exit area, P_{vacuum} is the vacuum pressure, and P_a is the ambient atmospheric pressure. Consequently, as the rocket gains altitude and atmospheric pressure reduces, thrust increases.

The weight of the rocket decreases as propellant is consumed:

$$W(t) = (m_{\text{dry}} + m_{\text{propellant}}(t)) \cdot g_0$$

where m_{dry} is the dry mass of the rocket and $m_{\text{propellant}}(t)$ is the time-dependent propellant mass. Thus, with a continuously changing mass, the dynamic thrust-to-weight ratio becomes:

$$\text{TWR}_{\text{dynamic}}(t) = \frac{T(t)}{m(t) \cdot g_0}$$

The analysis of $\text{TWR}_{\text{dynamic}}$ entails optimizing engine performance and fuel efficiency through the flight arc, maintaining structural integrity, and achieving desired velocity profiles.

Payload Capacity and Thrust-to-Weight Ratio Optimization

Determining the optimal thrust-to-weight ratio involves a trade-off between maximizing payload capacity and ensuring sufficient thrust to achieve trajectory objectives. Payload capacity is inherently linked to the amount of propellant and structural support, impacting the overall mass:

$$m_{\text{rocket}} = m_{\text{dry}} + m_{\text{propellant}} + m_{\text{payload}}$$

To optimize payload efficiency, the use of advanced propulsion systems and lightweight materials becomes pivotal. The design equation for an optimal TWR that accommodates a specified payload capacity can be expressed through a rearrangement and integration of mass flow rates in accordance with propulsion power budgets and specific impulse (I_{sp}):

$$\text{TWR}_{\text{optimal}} = \frac{T_{\text{efficient}}}{(m_{\text{dry}} + m_{\text{efficient propellant}}) \cdot g_0}$$

here, $m_{\text{efficient propellant}}$ is calculated from optimal usage strategies and propulsion efficiency. Approaching these analyses requires a firm foundation in propulsion thermodynamics and structural mechanics.

Applications in Different Phases of Flight

The thrust-to-weight ratio varies significantly through different phases of flight, from launch to stage separation to orbit insertion. Each phase demands precise calculations adjusting for vehicle mass variations and thrust configurations:
1. **Launch Phase**: High TWR ensures rapid clearance off the launch pad, mitigating adverse atmospheric effects. 2. **Ascent and Acceleration**: Moderate TWR accommodates controlled acceleration, optimizing propellant consumption rates. 3. **Orbit Insertion**: Targeted TWR aligns thrust with necessary velocity changes for achieving and maintaining orbital paths.

An integrated approach combines propulsion design, structural materials, and mission planning to ensure optimal thrust-to-weight ratios adapted to each phase.

Python Code Snippet

Below is a Python code snippet that encompasses the key computations related to thrust-to-weight ratio, including static and dynamic calculations, optimization techniques, and ascent considerations.

```
import numpy as np

def static_twr(thrust, mass, g0=9.81):
    '''
    Calculate static thrust-to-weight ratio.
    :param thrust: Total thrust produced by the rocket engines.
```

```
    :param mass: Total mass of the rocket.
    :param g0: Standard gravitational acceleration. Default is 9.81
    ↪  m/s^2.
    :return: Static thrust-to-weight ratio.
    '''
    return thrust / (mass * g0)

def dynamic_thrust(vacuum_thrust, exit_area, vacuum_pressure,
↪  ambient_pressure):
    '''
    Calculate the dynamic thrust during ascent.
    :param vacuum_thrust: Engine thrust under vacuum conditions.
    :param exit_area: Engine exit area.
    :param vacuum_pressure: Vacuum pressure.
    :param ambient_pressure: Ambient atmospheric pressure.
    :return: Thrust adjusted for altitude.
    '''
    return vacuum_thrust - exit_area * (vacuum_pressure -
↪  ambient_pressure)

def dynamic_weight(dry_mass, propellant_mass, g0=9.81):
    '''
    Calculate dynamic weight of the rocket as propellant is burned.
    :param dry_mass: Dry mass of the rocket.
    :param propellant_mass: Current mass of the propellant.
    :param g0: Standard gravitational acceleration. Default is 9.81
    ↪  m/s^2.
    :return: Dynamic weight.
    '''
    return (dry_mass + propellant_mass) * g0

def dynamic_twr(thrust, dry_mass, propellant_mass):
    '''
    Calculate dynamic thrust-to-weight ratio during ascent.
    :param thrust: Current thrust during ascent.
    :param dry_mass: Dry mass of the rocket.
    :param propellant_mass: Current mass of the propellant.
    :return: Dynamic thrust-to-weight ratio.
    '''
    weight = dynamic_weight(dry_mass, propellant_mass)
    return thrust / weight

def optimize_twr(vacuum_thrust, dry_mass, efficient_propellant_mass,
↪  g0=9.81):
    '''
    Calculate the optimal thrust-to-weight ratio based on efficient
    ↪  propellant usage.
    :param vacuum_thrust: Efficient thrust in vacuum conditions.
    :param dry_mass: Dry mass of the rocket.
    :param efficient_propellant_mass: Optimized mass of the
    ↪  propellant.
    :param g0: Standard gravitational acceleration.
    :return: Optimal thrust-to-weight ratio.
```

```
    return vacuum_thrust / ((dry_mass + efficient_propellant_mass) *
    ↪  g0)

# Example data for computations
thrust = 1e6  # in Newtons
mass = 1e5  # in kg

# Static TWR
static_ratio = static_twr(thrust, mass)
print("Static TWR:", static_ratio)

# Dynamic calculations along a sample ascent profile
vacuum_thrust = 1.2e6   # N
exit_area = 20.0   # m^2
vacuum_pressure = 0.0   # Pa
ambient_pressure = 101325   # Pa (at sea level)

thrust_during_ascent = dynamic_thrust(vacuum_thrust, exit_area,
    ↪  vacuum_pressure, ambient_pressure)
print("Thrust during ascent:", thrust_during_ascent)

dry_mass = 90000   # kg
propellant_mass = 10000   # kg
current_dynamic_twr = dynamic_twr(thrust_during_ascent, dry_mass,
    ↪  propellant_mass)
print("Dynamic TWR during ascent:", current_dynamic_twr)

# Optimization calculations
efficient_propellant_mass = 8000   # kg
optimal_twr = optimize_twr(vacuum_thrust, dry_mass,
    ↪  efficient_propellant_mass)
print("Optimal TWR:", optimal_twr)
```

This code defines several key functions for the computation and optimization of the thrust-to-weight ratio:

- `static_twr` function calculates the static thrust-to-weight ratio at lift-off.
- `dynamic_thrust` adjusts thrust for altitude-related atmospheric changes during ascent.
- `dynamic_weight` computes the dynamic weight as the rocket consumes propellant.
- `dynamic_twr` calculates the dynamic thrust-to-weight ratio as a function of changing thrust and mass.
- `optimize_twr` finds the optimal thrust-to-weight ratio based on efficient propellant usage.

The final block of code provides example calculations using dummy data to illustrate the application of these functions in rocket propulsion analysis.

Chapter 15

Engine Cycle Efficiency

Efficiency Metrics in Rocket Engine Cycles

Rocket engine efficiency is a paramount factor in aeronautics, facilitating the optimal conversion of chemical energy into kinetic energy. This efficiency is often categorized into thermal, propulsive, and overall system efficiency. The thermal efficiency, η_{thermal}, reflects how effectively the engine converts chemical energy from propellant combustion into kinetic energy and is given by:

$$\eta_{\text{thermal}} = \frac{T_s - T_e}{T_s}$$

where T_s is the stagnation temperature of the combustion chamber and T_e is the exhaust temperature. Propulsive efficiency, $\eta_{\text{propulsive}}$, expresses the effectiveness of converting exhaust velocity into useful thrust, defined as:

$$\eta_{\text{propulsive}} = \frac{2v_0}{v_{\text{exhaust}} + v_0}$$

Here, v_0 is the vehicle velocity and v_{exhaust} is the effective exhaust velocity. The cycle efficiency, η_{cycle}, amalgamates these metrics to quantify overall performance.

Staged Combustion Cycle Efficiency

Staged combustion engines, such as those in some liquid rocket engines, leverage full propellant combustion through divided processes. This involves preburner stages followed by main combustion, addressed by specific performance equations. The efficiency of staged combustion, η_staged, can be articulated as:

$$\eta_\text{staged} = \left(\frac{\delta H_c}{\delta H_p + \delta H_o} \right) \cdot \eta_\text{comb}$$

where δH_c is the enthalpy increase due to complete propellant combustion, δH_p and δH_o denote the specific enthalpy changes of the preburner fuel and oxidizer, respectively, and η_comb signifies the combustion efficiency at each stage.

Expander Cycle Efficiency

Expander cycle engines exploit the heating of cryogenic propellants through the warming of engine components for generating exhaust flow, which results in a different efficiency pattern. The expander cycle efficiency, η_expander, can be formulated using the relationship:

$$\eta_\text{expander} = \frac{W_\text{turb}}{\delta H_\text{total}}$$

Here, W_turb represents the work produced by the turbine as a result of propellant heating, and δH_total is the total enthalpy available from the combusted propellant. The expander cycle requires an intricate balance of thermal heat transfer and propellant dynamics to achieve peak system efficiencies.

Influence of Cycle Parameters on Efficiency

Efficiency metrics in both staged combustion and expander cycles are directly influenced by various system parameters, including chamber pressure, fuel-oxidizer ratios, and nozzle configuration. The influence of chamber pressure, P_c, on cycle efficiency is manifested via the propellant flow relation:

$$\dot{m}_f = \frac{P_c \cdot A_t}{c_*}$$

where $\dot{m}_f$ is the propellant flow rate, A_t is the throat area, and c_* is the characteristic velocity. An optimal fuel-oxidizer ratio, $(F/O)_{\text{opt}}$, contributes substantially to maximizing thermal and propulsive efficiencies, ensuring a Stoichiometric combustion process while maintaining structural integrity.

Mathematical Modeling and Simulation

Advanced mathematical modeling significantly enhances the ability to predict engine cycle efficiency under various operating conditions. Simulation tools often involve solving the governing equations of fluid dynamics, thermodynamics, and heat transfer numerically:

$$\frac{\partial}{\partial t}\int_V \rho\Phi\, dV + \oint_S \rho\Phi\vec{v}\cdot d\vec{A} = \int_V q_\Phi\, dV$$

where Φ represents the conserved quantity, ρ is the fluid density, and q_Φ encompasses the sources or sinks of Φ within volume V. Implementing these equations allows for a compressive evaluation of cycle performance, informing the design and optimization stages of rocket engine development.

Python Code Snippet

Below is a Python code snippet that includes calculations for rocket engine cycle efficiency, including thermal efficiency, propulsive efficiency, staged combustion cycle efficiency, expander cycle efficiency, and considering the influence of cycle parameters.

```
import numpy as np

def thermal_efficiency(T_s, T_e):
    '''
    Calculate the thermal efficiency of a rocket engine.
    :param T_s: Stagnation temperature of the combustion chamber.
    :param T_e: Exhaust temperature.
    :return: Thermal efficiency.
    '''
    return (T_s - T_e) / T_s

def propulsive_efficiency(v_0, v_exhaust):
    '''
    Calculate the propulsive efficiency of a rocket engine.
```

```python
    :param v_0: Vehicle velocity.
    :param v_exhaust: Effective exhaust velocity.
    :return: Propulsive efficiency.
    '''
    return (2 * v_0) / (v_exhaust + v_0)

def staged_combustion_efficiency(delta_Hc, delta_Hp, delta_Ho,
↪    eta_comb):
    '''
    Calculate the staged combustion cycle efficiency.
    :param delta_Hc: Enthalpy increase due to complete propellant
↪        combustion.
    :param delta_Hp: Specific enthalpy change of the preburner fuel.
    :param delta_Ho: Specific enthalpy change of the preburner
↪        oxidizer.
    :param eta_comb: Combustion efficiency at each stage.
    :return: Staged combustion efficiency.
    '''
    return (delta_Hc / (delta_Hp + delta_Ho)) * eta_comb

def expander_cycle_efficiency(W_turb, delta_H_total):
    '''
    Calculate the expander cycle efficiency.
    :param W_turb: Work produced by the turbine.
    :param delta_H_total: Total enthalpy available from combusted
↪        propellant.
    :return: Expander cycle efficiency.
    '''
    return W_turb / delta_H_total

def propellant_flow_rate(P_c, A_t, c_star):
    '''
    Calculate the propellant flow rate based on chamber pressure.
    :param P_c: Chamber pressure.
    :param A_t: Throat area.
    :param c_star: Characteristic velocity.
    :return: Propellant flow rate.
    '''
    return (P_c * A_t) / c_star

# Example parameters for demonstration
T_s = 3500          # K
T_e = 2500          # K
v_0 = 2500          # m/s
v_exhaust = 3000    # m/s
delta_Hc = 1e7      # J/kg
delta_Hp = 2e6      # J/kg
delta_Ho = 3e6      # J/kg
eta_comb = 0.95
W_turb = 5e6        # J
delta_H_total = 6e6 # J
P_c = 10e6          # Pa
A_t = 0.1           # m^2
```

```
c_star = 1500      # m/s

# Perform calculations
thermal_eff = thermal_efficiency(T_s, T_e)
propulsive_eff = propulsive_efficiency(v_0, v_exhaust)
staged_eff = staged_combustion_efficiency(delta_Hc, delta_Hp,
↪   delta_Ho, eta_comb)
expander_eff = expander_cycle_efficiency(W_turb, delta_H_total)
flow_rate = propellant_flow_rate(P_c, A_t, c_star)

print("Thermal Efficiency:", thermal_eff)
print("Propulsive Efficiency:", propulsive_eff)
print("Staged Combustion Efficiency:", staged_eff)
print("Expander Cycle Efficiency:", expander_eff)
print("Propellant Flow Rate:", flow_rate)
```

This code defines several key functions necessary for evaluating engine cycle efficiencies, including:

- `thermal_efficiency` computes how effectively the engine converts chemical energy into kinetic energy.

- `propulsive_efficiency` measures the efficiency of converting exhaust velocity into thrust.

- `staged_combustion_efficiency` assesses the efficiency in staged combustion cycles.

- `expander_cycle_efficiency` calculates the efficiency of expander cycles.

- `propellant_flow_rate` evaluates the flow rate of propellant based on chamber properties.

Using hypothetical parameter values, the final block demonstrates these computations, showing how different efficiencies can be quantified for given engine conditions.

Chapter 16

Pressure and Temperature Calculations

Thermodynamic Principles in Rocket Engines

The operation of rocket propulsion systems is rooted in thermodynamic principles. Understanding the variations of pressure and temperature within rocket engines is fundamental to ensuring their structural integrity and operational efficiency. The fundamental equation of state for an ideal gas is often employed:

$$PV = nRT$$

where P denotes pressure, V is volume, n is the number of moles, R is the ideal gas constant, and T is the temperature. In the context of rocket engines, deviations from ideal gas behavior are considered through corrections such as compressibility factors.

Isentropic Flow and Engine Performance

Isentropic flow conditions, where entropy remains constant, serve as a critical reference for analyzing performance. The relationship between pressure and temperature in an isentropic process is described by:

$$\left(\frac{T_2}{T_1}\right) = \left(\frac{P_2}{P_1}\right)^{\frac{\gamma-1}{\gamma}}$$

where P_1 and P_2 are initial and final pressures, T_1 and T_2 are initial and final temperatures, and γ is the specific heat ratio $\left(\frac{C_p}{C_v}\right)$.

The temperature rise in a combustion chamber can be approximated by:

$$\Delta T = \frac{q}{C_p}$$

where ΔT is the temperature change, q is the heat added per unit mass, and C_p is the specific heat at constant pressure.

Pressure and Temperature in Nozzle Design

The design of rocket nozzles involves careful consideration of pressure and temperature changes. The critical throat pressure P^* is determined as:

$$P^* = P_0 \left(\frac{2}{\gamma+1}\right)^{\frac{\gamma}{\gamma-1}}$$

where P_0 is the stagnation pressure. The corresponding temperature T^* at the nozzle throat is calculated by:

$$T^* = T_0 \left(\frac{2}{\gamma+1}\right)$$

The Mach number M_e at the nozzle exit relates to temperature and pressure through:

$$M_e^2 = \frac{2}{\gamma-1}\left(\left(\frac{P_0}{P_e}\right)^{\frac{\gamma-1}{\gamma}} - 1\right)$$

where P_e is the exhaust pressure. Accurate determination of these parameters is crucial for optimizing thrust and efficiency.

Influence of Combustion on Pressure and Temperature

The combustion process in rocket engines significantly influences pressure and temperature profiles. The maximum pressure, often referred to as chamber pressure P_c, depends on combustion dynamics:

$$P_c = P_0 + \frac{RdT}{M_w}$$

where dT is the differential temperature across the chamber, R is the universal gas constant, and M_w is the molecular weight of the combustion products.

Combustion efficiency and heat release rates directly impact P_c and temperature distribution, requiring robust control and monitoring to maintain operational stability.

Structural Material Considerations for Pressure and Temperature

Rocket engine designs must accommodate high-pressure and high-temperature environments. The material selection process is guided by the maximum allowable stress, calculated using:

$$\sigma_{\max} = \frac{P_c \cdot A}{\text{FS}}$$

where $\sigma_{\max}$ is the maximum stress, A is the cross-sectional area, and FS is the factor of safety.

Thermal expansion influences are accounted for through the coefficient of thermal expansion α:

$$\Delta L = \alpha L_0 \Delta T$$

where ΔL is the change in length, L_0 is the original length, and ΔT is the temperature change. Understanding how materials respond to these dynamic conditions is essential for the integrity of rocket engine systems.

Python Code Snippet

Below is a Python code snippet that encompasses the core computational elements of pressure and temperature calculations within rocket engines, including calculations for isentropic flow, nozzle design, and material considerations.

```python
import numpy as np

def ideal_gas_pressure(n, V, R, T):
    '''
    Calculate pressure from the ideal gas law.
    :param n: Number of moles.
    :param V: Volume.
    :param R: Ideal gas constant.
    :param T: Temperature.
    :return: Pressure.
    '''
    return (n * R * T) / V

def isentropic_temperature_ratio(P1, P2, gamma):
    '''
    Calculate the temperature ratio for an isentropic process.
    :param P1: Initial pressure.
    :param P2: Final pressure.
    :param gamma: Specific heat ratio.
    :return: Temperature ratio.
    '''
    return (P2 / P1) ** ((gamma - 1) / gamma)

def combustion_temperature_increase(q, Cp):
    '''
    Calculate the temperature increase in a combustion process.
    :param q: Heat added per unit mass.
    :param Cp: Specific heat at constant pressure.
    :return: Temperature change.
    '''
    return q / Cp

def critical_throat_conditions(P0, T0, gamma):
    '''
    Determine critical pressure and temperature at the nozzle
        throat.
    :param P0: Stagnation pressure.
    :param T0: Stagnation temperature.
    :param gamma: Specific heat ratio.
    :return: Critical throat pressure and temperature.
    '''
    P_star = P0 * (2 / (gamma + 1)) ** (gamma / (gamma - 1))
    T_star = T0 * (2 / (gamma + 1))
    return P_star, T_star
```

```python
def exit_mach_number(P0, Pe, gamma):
    '''
    Calculate the Mach number at the nozzle exit.
    :param P0: Stagnation pressure.
    :param Pe: Exhaust pressure.
    :param gamma: Specific heat ratio.
    :return: Mach number at the exit.
    '''
    return np.sqrt((2 / (gamma - 1)) * ((P0 / Pe) ** ((gamma - 1) /
    ↪   gamma) - 1))

def chamber_pressure(P0, dT, R, Mw):
    '''
    Calculate the chamber pressure considering combustion.
    :param P0: Initial pressure.
    :param dT: Differential temperature across the chamber.
    :param R: Universal gas constant.
    :param Mw: Molecular weight of combustion products.
    :return: Chamber pressure.
    '''
    return P0 + (R * dT / Mw)

def max_allowable_stress(Pc, A, FS):
    '''
    Calculate the maximum allowable stress for material selection.
    :param Pc: Chamber pressure.
    :param A: Cross-sectional area.
    :param FS: Factor of safety.
    :return: Maximum allowable stress.
    '''
    return Pc * A / FS

def thermal_expansion(alpha, L0, dT):
    '''
    Calculate change in length due to thermal expansion.
    :param alpha: Coefficient of thermal expansion.
    :param L0: Original length.
    :param dT: Temperature change.
    :return: Change in length.
    '''
    return alpha * L0 * dT

# Constants and example inputs for demonstration
n = 1.0      # moles
V = 22.4     # liters
R = 8.314    # J/(mol*K)
T = 300.0    # Kelvin
P1 = 101325  # Pa
P2 = 202650  # Pa
gamma = 1.4
q = 50000    # J/kg
Cp = 1005    # J/(kg*K)
```

```
P0 = 3e6      # Pa
T0 = 3500.0   # Kelvin
Pe = 101325   # Pa
dT = 250.0    # K
Mw = 0.02896  # kg/mol
A = 0.01      # m^2
FS = 1.5
alpha = 1e-5  # 1/K
L0 = 1.0      # m

# Calculations
pressure = ideal_gas_pressure(n, V, R, T)
temp_ratio = isentropic_temperature_ratio(P1, P2, gamma)
temp_increase = combustion_temperature_increase(q, Cp)
P_star, T_star = critical_throat_conditions(P0, T0, gamma)
Me = exit_mach_number(P0, Pe, gamma)
Pc = chamber_pressure(P0, dT, R, Mw)
sigma_max = max_allowable_stress(Pc, A, FS)
length_change = thermal_expansion(alpha, L0, dT)

print("Calculated Pressure:", pressure)
print("Isentropic Temperature Ratio:", temp_ratio)
print("Temperature Increase due to Combustion:", temp_increase)
print("Critical Throat Conditions (P*, T*):", (P_star, T_star))
print("Exit Mach Number:", Me)
print("Chamber Pressure:", Pc)
print("Maximum Allowable Stress:", sigma_max)
print("Length Change due to Thermal Expansion:", length_change)
```

This code defines several key functions necessary for the analysis of rocket engine pressure and temperature dynamics:

- ideal_gas_pressure function calculates pressure based on the ideal gas law.

- isentropic_temperature_ratio evaluates the temperature ratio for isentropic processes.

- combustion_temperature_increase determines temperature changes from combustion heat.

- critical_throat_conditions provides critical pressure and temperature in the nozzle.

- exit_mach_number computes the Mach number at nozzle exit conditions.

- chamber_pressure estimates the chamber pressure affected by combustion dynamics.

- `max_allowable_stress` evaluates maximum stress allowable for material selection.

- `thermal_expansion` calculates changes in material length due to thermal expansion.

The final block of code demonstrates example calculations using specified constants and inputs.

Chapter 17

Empirical Equations for Exhaust Velocity

Introduction to Exhaust Velocity Calculations

Exhaust velocity, a pivotal parameter in rocket propulsion, fundamentally dictates the thrust generated by rocket engines. It is influenced heavily by the choice of propellant and the design of the nozzle. The empirical determination of exhaust velocity for various combinations is essential to optimize propulsion performance. The exhaust velocity v_e is obtained as:

$$v_e = C^* \times \eta$$

where C^* is the characteristic velocity and η is the nozzle efficiency factor, indicative of losses from non-ideal flow and thermodynamic inefficiencies.

Characteristic Velocity Consideration

The characteristic velocity, C^*, is vital in assessing the efficiency of propellant combustion. It is defined by:

$$C^* = \frac{P_c A_t}{\dot{m}}$$

where P_c represents the chamber pressure, A_t the throat area of the nozzle, and $\dot{m}$ the mass flow rate of the exhaust gases. This semi-empirical relation is influenced by chemical properties and combustion efficiency, as determined by:

$$C^* = \frac{\sqrt{RT_c}}{\gamma}\left(\frac{2}{\gamma+1}\right)^{(\gamma+1)/(2(\gamma-1))}$$

where R represents the specific gas constant, T_c the combustion chamber temperature, and γ the specific heat ratio of the exhaust gases.

Nozzle Efficiency and Design Implications

Nozzle efficiency, η, modifies the ideal exhaust velocity predictions to account for real-world phenomena such as boundary layer separation and geometric constraints. Efficiency adjustments are carried out using empirical loss co-efficients specific to propellant type and nozzle design. The effective exhaust velocity v_e becomes:

$$v_e = \eta\sqrt{\frac{2\gamma RT_c}{\gamma-1}\left[1-\left(\frac{P_e}{P_c}\right)^{(\gamma-1)/\gamma}\right]}$$

where P_e is the exit pressure. Deviations from the isentropic flow model are captured by η.

1 Empirical Correction Factors

The application of empirical correction factors is essential to fine-tune the exhaust velocity for variations in nozzle Mach number and geometry. Utilizing experimental data, correlations are developed to quantify efficiency losses. The correction factor K is established such that:

$$v_e = K \times \sqrt{\frac{2\gamma RT_c}{\gamma-1}\left[1-\left(\frac{P_e}{P_c}\right)^{(\gamma-1)/\gamma}\right]}$$

Variable Propellant Combinations

The choice of propellant combination plays a pivotal role in determining exhaust velocity and therefore thrust. Each propellant combination possesses inherent properties which affect chamber pressure P_c, combustion temperature T_c, and specific gas constant R. These factors are incorporated into the empirical relations:

$$v_e = F_{prop}\sqrt{\frac{2\gamma_{prop}R_{prop}T_c}{\gamma_{prop}-1}\left[1-\left(\frac{P_e}{P_c}\right)^{(\gamma_{prop}-1)/\gamma_{prop}}\right]}$$

where F_{prop} is an empirical factor unique to each propellant combination.

1 Propellant Performance Indices

A performance index I_{prop} is often used to rate propellant combinations for specific applications, representing the relative efficiency enhancements they offer when combined with different nozzle designs. The empirical relation:

$$I_{prop} = \frac{v_{e,prop}}{v_{e,ref}}$$

where $v_{e,prop}$ is the exhaust velocity with the propellant in question and $v_{e,ref}$ a reference velocity, provides an insightful metric for performance optimization.

Conclusion on Exhaust Velocity Optimization

Empirical equations forge a path for accurate and efficient prediction of exhaust velocity, essential for maximizing rocket engine efficiency. The intricate interplay of propellant properties, nozzle design, and empirical corrections is captured within these equations, allowing aerospace engineers to strategically tailor propulsion systems to meet specified performance criteria.

Python Code Snippet

Below is a Python code snippet that encompasses the core computational elements necessary for the calculations related to exhaust velocity, characteristic velocity, and nozzle efficiency as discussed in this chapter.

```python
import numpy as np

def characteristic_velocity(P_c, A_t, m_dot, R, T_c, gamma):
    '''
    Calculate the characteristic velocity.
    :param P_c: Chamber pressure in Pascals.
    :param A_t: Throat area in square meters.
    :param m_dot: Mass flow rate in kg/s.
    :param R: Specific gas constant in J/(kg*K).
    :param T_c: Combustion chamber temperature in Kelvin.
    :param gamma: Specific heat ratio.
    :return: Characteristic velocity in m/s.
    '''
    C_star = (P_c * A_t) / m_dot
    return C_star, np.sqrt(R * T_c) / gamma * (2 / (gamma + 1)) ** ((gamma + 1) / (2 * (gamma - 1)))

def exhaust_velocity(C_star, eta, R, T_c, gamma, P_e, P_c):
    '''
    Calculate the effective exhaust velocity.
    :param C_star: Characteristic velocity.
    :param eta: Nozzle efficiency.
    :param R: Specific gas constant in J/(kg*K).
    :param T_c: Combustion chamber temperature in Kelvin.
    :param gamma: Specific heat ratio.
    :param P_e: Exit pressure in Pascals.
    :param P_c: Chamber pressure in Pascals.
    :return: Effective exhaust velocity in m/s.
    '''
    v_e = eta * np.sqrt(2 * gamma * R * T_c / (gamma - 1) * (1 - (P_e / P_c) ** ((gamma - 1) / gamma)))
    return v_e

def empirical_corrected_velocity(R, T_c, gamma, P_e, P_c, K):
    '''
    Calculate the exhaust velocity using empirical correction
        factors.
    :param R: Specific gas constant in J/(kg*K).
    :param T_c: Combustion chamber temperature in Kelvin.
    :param gamma: Specific heat ratio.
    :param P_e: Exit pressure in Pascals.
    :param P_c: Chamber pressure in Pascals.
    :param K: Empirical correction factor.
    :return: Corrected exhaust velocity in m/s.
    '''
    v_e_corrected = K * np.sqrt(2 * gamma * R * T_c / (gamma - 1) * (1 - (P_e / P_c) ** ((gamma - 1) / gamma)))
    return v_e_corrected

# Example data
P_c = 3.5e6  # Chamber pressure in Pascals
A_t = 0.02   # Throat area in square meters
```

```
m_dot = 25.0   # Mass flow rate in kg/s
R = 287.0      # Specific gas constant for air in J/(kg*K)
T_c = 3500.0   # Combustion chamber temperature in Kelvin
gamma = 1.4    # Specific heat ratio
eta = 0.95     # Nozzle efficiency
P_e = 1.0e5    # Exit pressure in Pascals
K = 0.98       # Empirical correction factor

# Calculations
C_star, C_star_ideal = characteristic_velocity(P_c, A_t, m_dot, R,
↪   T_c, gamma)
v_e = exhaust_velocity(C_star, eta, R, T_c, gamma, P_e, P_c)
v_e_corrected = empirical_corrected_velocity(R, T_c, gamma, P_e,
↪   P_c, K)

print("Characteristic Velocity:", C_star, "m/s (Ideal:",
↪   C_star_ideal, "m/s)")
print("Effective Exhaust Velocity:", v_e, "m/s")
print("Empirically Corrected Exhaust Velocity:", v_e_corrected,
↪   "m/s")
```

This code defines several key functions necessary for calculating rocket propulsion parameters:

- characteristic_velocity calculates the characteristic velocity using both its empirical form and the ideal thermodynamic form.

- exhaust_velocity computes the effective exhaust velocity using nozzle efficiency and environmental pressure conditions.

- empirical_corrected_velocity calculates exhaust velocity with adjustments based on empirical correction factors, capturing the practical deviations from ideal predictions.

The provided example performs calculations with exemplary data, yielding values essential for propulsion analysis.

Chapter 18

Aerodynamics in Rocket Design

Introduction to Aerodynamic Forces

Rocket aerodynamics is pivotal in influencing the trajectory and stability of rockets during atmospheric flight. Aerodynamic forces arise from pressure differentials and viscous interactions as a rocket moves through a fluid medium. Formally, the aerodynamic force $\mathbf{F}_a$ can be expressed as:

$$\mathbf{F}_a = \frac{1}{2}\rho V^2 S C_f$$

where ρ denotes the air density, V the velocity relative to the air, S the reference area, and C_f the coefficient of aerodynamic force, which encapsulates the effects of shape, inclination, and surface roughness.

Pressure Drag and Lift Components

The drag force (D) and lift force (L) are two fundamental components of aerodynamic force. The drag force acts parallel to the oncoming flow direction, while the lift force acts perpendicular to it. These can be mathematically represented as:

$$D = \frac{1}{2}\rho V^2 S C_D$$

$$L = \frac{1}{2}\rho V^2 S C_L$$

where C_D and C_L are the drag and lift coefficients, respectively, determined through empirical data or computational simulations.

Mathematical Modeling of Rocket Shape

The rocket's shape significantly impacts aerodynamic efficiency and can be optimized using mathematical modeling. Considerations of slender bodies lead to employing potential flow theory, where the pressure distribution around the rocket is a function of its geometry and angle of attack α. The relation is expressed as:

$$C_p = C_{p,0} + m\alpha$$

where C_p is the pressure coefficient, $C_{p,0}$ is the pressure coefficient at zero angle of attack, and m represents the slope of the lifting line theory.

Optimization Techniques in Rocket Aerodynamics

Aerodynamic optimization requires iterative methods such as gradient-based and evolutionary algorithms. The objective function often involves minimizing drag while ensuring stability and control, which mathematically can be formulated as:

$$\text{minimize } J(\mathbf{x}) = \int_S C_D(\mathbf{x}, s)\, ds$$

subject to constraints related to the lift, $C_L(\mathbf{x}) \geq C_{L,\min}$, and structural limitations. Here, $\mathbf{x}$ represents the design variables, such as curvature and surface angle.

1 Empirical Corrections and Fluid-Structure Interaction

The interaction between the aerodynamic loads and the structural dynamics of rockets requires advanced modeling techniques, particularly in the transonic and supersonic regimes. Empirical corrections to the CFD models incorporate these phenomena with:

$$F_i = M\frac{\mathrm{d}^2 u_i}{\mathrm{d}t^2} + C\frac{\mathrm{d}u_i}{\mathrm{d}t} + Ku_i = f_a\left(\frac{\partial u_i}{\partial x_j}, \rho, M, V\right)$$

where M is the mass matrix, C is the damping matrix, K is the stiffness matrix, u_i is the displacement vector, and f_a represents the aerodynamic forcing function as per the displacement vector gradients and flow conditions. These inform the coupled numeric simulations requisite for high-fidelity aerodynamic predictions.

Python Code Snippet

Below is a Python code snippet that encompasses the core calculations for aerodynamic forces, drag and lift analysis, optimization techniques, and fluid-structure interaction modeling in rocket design.

```
import numpy as np
from scipy.optimize import minimize

def aerodynamic_force(rho, V, S, C_f):
    '''
    Compute the aerodynamic force.
    :param rho: Air density.
    :param V: Velocity relative to the air.
    :param S: Reference area.
    :param C_f: Coefficient of aerodynamic force.
    :return: Aerodynamic force.
    '''
    return 0.5 * rho * V**2 * S * C_f

def drag_lift_forces(rho, V, S, C_D, C_L):
    '''
    Compute the drag and lift forces.
    :param rho: Air density.
    :param V: Velocity.
    :param S: Reference area.
    :param C_D: Drag coefficient.
    :param C_L: Lift coefficient.
    :return: Drag and lift forces.
    '''
    D = 0.5 * rho * V**2 * S * C_D
    L = 0.5 * rho * V**2 * S * C_L
    return D, L

def pressure_coefficient(C_p0, m, alpha):
    '''
    Calculate pressure coefficient based on angle of attack.
```

```
    :param C_p0: Pressure coefficient at zero angle of attack.
    :param m: Slope of lift line theory.
    :param alpha: Angle of attack.
    :return: Pressure coefficient.
    '''
    return C_p0 + m * alpha

def aerodynamic_optimization(C_D, C_L_min, constraints):
    '''
    Optimize rocket design for aerodynamic efficiency.
    :param C_D: Drag coefficient as function of design variables.
    :param C_L_min: Minimum lift coefficient.
    :param constraints: Design constraints for the optimization
    ↪ process.
    :return: Optimized design variables.
    '''
    objective_function = lambda x: np.sum(C_D(x))
    result = minimize(objective_function, x0,
    ↪ constraints=constraints)
    return result.x

def fluid_structure_interaction_analysis(M, C, K, f_a, u_i):
    '''
    Analyze fluid-structure interaction using aerodynamics and
    ↪ structural dynamics.
    :param M: Mass matrix.
    :param C: Damping matrix.
    :param K: Stiffness matrix.
    :param f_a: Aerodynamic force function.
    :param u_i: Initial displacement vector.
    :return: New displacement vector and acceleration.
    '''
    # Example of Newton-Raphson method for FSI solution
    # Simplified for illustrative purposes
    FSI_solution = np.linalg.solve(
        M + C + K,
        f_a(u_i) - np.dot(C, u_i) - np.dot(K, u_i)
    )
    return FSI_solution

# Example Inputs
rho = 1.225    # kg/m^3
V = 340.29     # m/s (speed of sound at sea level)
S = 10.0       # m^2
C_f = 0.04
C_D = np.array([0.02, 0.03, 0.04])    # Example drag coefficients for
↪ variable surfaces
C_L = 0.5
C_p0 = 0.1
m = 0.05
alpha = 5.0    # degrees

# Computation
```

```
aero_force = aerodynamic_force(rho, V, S, C_f)
D, L = drag_lift_forces(rho, V, S, C_D[0], C_L)
pressure_co = pressure_coefficient(C_p0, m, alpha)

# Constraints and Optimization
constraints = [{'type': 'ineq', 'fun': lambda x: x[0] - C_L}]
optimized_design = aerodynamic_optimization(lambda x: x**2, C_L,
↪   constraints)

print("Aerodynamic Force:", aero_force)
print("Drag Force:", D, "Lift Force:", L)
print("Pressure Coefficient:", pressure_co)
print("Optimized Design:", optimized_design)
```

This code provides a comprehensive Python implementation for calculating and optimizing aerodynamic forces in rocket design:

- `aerodynamic_force` calculates the total aerodynamic force on a rocket.

- `drag_lift_forces` computes individual drag and lift forces based on specified coefficients.

- `pressure_coefficient` determines the pressure coefficient from angle of attack using linear relations.

- `aerodynamic_optimization` is a procedure for optimizing rocket design by minimizing drag whilst meeting lift constraints.

- `fluid_structure_interaction_analysis` showcases a simplified analysis of aerodynamic loads coupled with structural dynamics.

These examples underscore how Python can be used to solve and optimize complex equations in aerospace engineering, providing high-level capabilities for simulation and design processes.

Chapter 19

Drag Force Calculations

Fundamentals of Drag Force

The drag force is an essential consideration in aerospace engineering, particularly during the ascent and re-entry phases of rocket flight. It is defined as the force opposing the relative motion of a body through a fluid and is expressed mathematically as:

$$D = \frac{1}{2}\rho V^2 S C_D \quad (19.1)$$

where D denotes the drag force, ρ represents the fluid density, V is the velocity relative to the fluid, S is the reference area, and C_D is the drag coefficient. Drag force analysis requires a precise understanding of these variables and the interactions between them.

Components of Drag

Airflow over a rocket during its flight can be disaggregated into various drag components, including pressure (form) drag, skin friction drag, and wave drag, especially at supersonic speeds.

1 Pressure Drag

Pressure drag, also known as form drag, arises due to the pressure differential over the rocket's surface in regions of flow separation.

This is modeled using the pressure coefficient C_p, which relates surface pressure to the free-stream dynamic pressure:

$$C_p = \frac{p - p_\infty}{\frac{1}{2}\rho V^2} \tag{19.2}$$

where p is the local pressure and p_∞ is the free-stream pressure.

2 Skin Friction Drag

Skin friction drag results from viscous shear forces acting on the rocket's surface and is quantitatively described by the friction coefficient:

$$C_f = \left(\frac{\tau_w}{\frac{1}{2}\rho V^2}\right) \tag{19.3}$$

Here, τ_w is the wall shear stress.

3 Wave Drag

Wave drag is significant at supersonic speeds and occurs due to shock waves formation, being characterized by the wave drag coefficient $C_{D,w}$. This demands analysis via supersonic wave theory for precise estimation.

Advanced Computational Techniques

The accurate prediction of drag forces is feasible through computational fluid dynamics (CFD), facilitating the resolution of complex flow fields around the rocket's geometry during high velocity phases. The governing equations of CFD, primarily the Navier-Stokes equations, are solved numerically:

$$\frac{\partial \rho}{\partial t} + \nabla \cdot (\rho \mathbf{V}) = 0 \tag{19.4}$$

$$\frac{\partial (\rho \mathbf{V})}{\partial t} + \nabla \cdot (\rho \mathbf{V}\mathbf{V}) = -\nabla p + \nabla \cdot \mathbf{T} \tag{19.5}$$

where $\mathbf{T}$ is the viscous stress tensor, requiring discretization techniques such as finite volume methods for solvability.

Minimization of Drag Forces

The minimization of drag during rocket flight is critical for performance optimization and can be approached through various design modifications and strategies.

1 Streamlining and Shape Optimization

The reduction of drag forces through shape optimization involves tailoring the rocket's geometry to minimize form drag. The application of slender body theory assists in configuring the optimal profile that delays flow separation:

$$L/D = f\left(\frac{l}{d}, \alpha\right) \tag{19.6}$$

where L and D are longitudinal and diameter dimensions of the rocket, respectively, and α is the angle of attack.

2 Re-entry Heating and Ablative Techniques

During re-entry, managing convective heat transfer and resultant drag is vital. The Stanton number St, a dimensionless heat transfer coefficient, aids in determining appropriate heat shield thickness:

$$St = \frac{h}{\rho V C_p} \tag{19.7}$$

where h is the heat transfer coefficient and C_p is the specific heat.

Empirical and Experimental Methods

Wind tunnel testing and flight data analysis remain integral to validating theoretical and computational models for drag prediction. The comprehensive experimental analysis where Reynolds numbers are matched with full-scale conditions assist in drag extrapolation.

Empirical correlations developed from such experiments refine C_D values, accounting for non-idealities present in real-world conditions.

The expert application of these methodologies culminates in the accurate estimation and reduction of drag forces, enhancing the operational efficacy of rocket systems during critical flight phases.

Python Code Snippet

Below is a Python code snippet that encompasses the core computational elements for drag force calculations, including the formulas for pressure drag, skin friction drag, wave drag, and streamlining strategies using computational fluid dynamics (CFD).

```python
import numpy as np

def calculate_drag_force(rho, V, S, C_D):
    '''
    Calculate the drag force experienced by the rocket.
    :param rho: Fluid density.
    :param V: Velocity relative to the fluid.
    :param S: Reference area.
    :param C_D: Drag coefficient.
    :return: Drag force value.
    '''
    return 0.5 * rho * V**2 * S * C_D

def pressure_drag(C_p, p, p_inf, rho, V):
    '''
    Calculate pressure drag using the pressure coefficient.
    :param C_p: Pressure coefficient.
    :param p: Local pressure.
    :param p_inf: Free-stream pressure.
    :param rho: Fluid density.
    :param V: Velocity relative to the fluid.
    :return: Pressure drag force.
    '''
    dynamic_pressure = 0.5 * rho * V**2
    return C_p * (p - p_inf) / dynamic_pressure

def skin_friction_drag(tau_w, rho, V):
    '''
    Calculate skin friction drag.
    :param tau_w: Wall shear stress.
    :param rho: Fluid density.
    :param V: Velocity relative to the fluid.
    :return: Skin friction drag force.
    '''
    dynamic_pressure = 0.5 * rho * V**2
    return tau_w / dynamic_pressure

def wave_drag(C_Dw, M):
    '''
    Calculate wave drag for supersonic speeds.
    :param C_Dw: Wave drag coefficient.
    :param M: Mach number.
    :return: Wave drag force.
    '''
```

```python
    # Placeholder for wave drag calculations
    return C_Dw * M**2

def streamline_optimization(l, d, alpha):
    '''
    Calculate length-to-diameter ratio optimization for drag
    ↪ minimization.
    :param l: Rocket length.
    :param d: Rocket diameter.
    :param alpha: Angle of attack.
    :return: Optimized L/D ratio.
    '''
    # This function can be enhanced with slender body theory
    ↪ calculations
    return l / d  # Simplified for demonstration

# Example calculations
rho = 1.225  # kg/m^3, density at sea level
V = 300  # m/s, velocity
S = 10  # m^2, reference area
C_D = 0.5  # Drag coefficient

drag_force = calculate_drag_force(rho, V, S, C_D)
print("Drag Force:", drag_force)

C_p = 0.2  # Example pressure coefficient
p = 101325  # Pa, local pressure
p_inf = 100000  # Pa, free-stream pressure
pressure_drag_force = pressure_drag(C_p, p, p_inf, rho, V)
print("Pressure Drag Force:", pressure_drag_force)

tau_w = 0.05  # Example wall shear stress in N/m^2
skin_friction_force = skin_friction_drag(tau_w, rho, V)
print("Skin Friction Drag Force:", skin_friction_force)

C_Dw = 0.1  # Example wave drag coefficient
M = 2  # Mach number for supersonic speed
wave_drag_force = wave_drag(C_Dw, M)
print("Wave Drag Force:", wave_drag_force)

l = 30  # m, rocket length
d = 3  # m, rocket diameter
alpha = 5  # degrees, angle of attack
L_D_ratio = streamline_optimization(l, d, alpha)
print("Optimized L/D Ratio:", L_D_ratio)
```

This code defines several key functions necessary for estimating and optimizing drag forces during rocket flights:

- `calculate_drag_force` computes the overall drag force using basic parameters such as fluid density, velocity, reference area, and drag coefficient.

- `pressure_drag` calculates the pressure drag based on the pressure coefficient model.

- `skin_friction_drag` evaluates skin friction drag via wall shear stress.

- `wave_drag` provides a framework for estimating wave drag at supersonic speeds.

- `streamline_optimization` showcases a simplified approach to optimizing the rocket's length-to-diameter ratio for drag reduction.

The final block of code demonstrates example calculations with assumed values, illustrating the application of these computational methods.

Chapter 20

Rocket Stability and Control

Stability Criteria and Equations

Rocket stability is a fundamental aspect of aerospace engineering, determining the vehicle's ability to maintain a desired trajectory or attitude. The stability criteria are typically analyzed in terms of static stability and dynamic stability. Static stability addresses the initial response of the rocket to perturbations, while dynamic stability concerns the behavior over time.

For static stability, the rocket's center of pressure must be aft of its center of gravity. Mathematically, this condition can be expressed as:

$$\Delta x_{cg-cp} = x_{cg} - x_{cp} > 0 \qquad (20.1)$$

where x_{cg} is the position of the center of gravity, and x_{cp} is the position of the center of pressure.

Dynamic stability can be assessed using the characteristic equation of the system, given by the determinant of the system matrix related to the rocket's linearized equations of motion:

$$\det(\mathbf{A} - \lambda \mathbf{I}) = 0 \qquad (20.2)$$

where $\mathbf{A}$ is the system matrix and λ represents the eigenvalues. The real parts of all eigenvalues must be negative to ensure stability.

Mathematical Models for Control Systems

Control systems in rocketry serve to maintain or alter the vehicle's flight path and attitude. The governing equations for control systems are typically represented as state-space models, defined as:

$$\dot{\mathbf{x}} = \mathbf{A}\mathbf{x} + \mathbf{B}\mathbf{u} \tag{20.3}$$

$$\mathbf{y} = \mathbf{C}\mathbf{x} + \mathbf{D}\mathbf{u} \tag{20.4}$$

where $\mathbf{x}$ represents the state vector, $\mathbf{u}$ is the control input vector, $\mathbf{y}$ is the output vector, and $\mathbf{A}$, $\mathbf{B}$, $\mathbf{C}$, and $\mathbf{D}$ are matrices defining the system dynamics and inputs/outputs.

Guidance System Equations and Feedback Control

Rocket guidance systems rely on feedback control mechanisms to adjust the thrust vector and navigation paths. Commonly, Proportional-Derivative-Integral (PID) controllers are implemented, with the control law expressed as:

$$u(t) = K_p e(t) + K_i \int e(t)\, dt + K_d \frac{d}{dt} e(t) \tag{20.5}$$

where $e(t)$ is the error signal, and K_p, K_i, and K_d are the proportional, integral, and derivative gains respectively.

The stability margin of a PID controller can be quantified through the Nyquist or Bode plot analysis, examining phase and gain margins.

Rotational Dynamics and Moment Equations

Rotational dynamics play a critical role in rocket stability and control, especially characterized by the Euler equations for rigid body motion:

$$\frac{d}{dt}(\mathbf{I}\boldsymbol{\omega}) + \boldsymbol{\omega} \times (\mathbf{I}\boldsymbol{\omega}) = \mathbf{M} \tag{20.6}$$

Here, $\mathbf{I}$ denotes the inertia tensor, $\boldsymbol{\omega}$ is the angular velocity vector, and $\mathbf{M}$ represents the moment vector applied to the body.

Attitude maneuvers can be executed by manipulating the moment applied through mechanisms such as gimbaled thrust or reaction control systems, governed by these fundamental equations.

Trajectory Optimization and Stability Analysis

Optimizing rocket trajectories involves calculating the most efficient path from launch to the final orbit, often using optimal control theory. The cost function J is minimized subject to constraints defined by the rocket's dynamic model:

$$J = \int_0^T L(\mathbf{x}, \mathbf{u}, t)\, dt \tag{20.7}$$

subject to the dynamics:

$$\dot{\mathbf{x}}(t) = f(\mathbf{x}(t), \mathbf{u}(t), t) \tag{20.8}$$

and boundary conditions at initial and terminal states.

The solutions of this optimization problem provide the necessary control inputs and trajectory states, ensuring stability and efficiency throughout the mission profile.

Python Code Snippet

Below is a Python code snippet that encompasses key computational elements related to rocket stability and control, including static stability evaluation, system dynamics, and PID control implementation for trajectory optimization and guidance.

```python
import numpy as np
from scipy.integrate import solve_ivp
from scipy.linalg import eigvals
import matplotlib.pyplot as plt

# Static Stability Function
def is_statically_stable(x_cg, x_cp):
    '''
    Determines static stability of the rocket.
    :param x_cg: Position of the center of gravity
    :param x_cp: Position of the center of pressure
    :return: Boolean indicating stability
    '''
```

```python
    return (x_cg - x_cp) > 0

# Dynamic Stability Function
def is_dynamically_stable(A):
    '''
    Determines dynamic stability by checking eigenvalues of system
    ↪ matrix.
    :param A: System matrix
    :return: Boolean indicating stability
    '''
    eigenvalues = eigvals(A)
    return np.all(np.real(eigenvalues) < 0)

# PID Controller
class PIDController:
    def __init__(self, Kp, Ki, Kd):
        self.Kp = Kp
        self.Ki = Ki
        self.Kd = Kd
        self.integral = 0
        self.previous_error = 0

    def update(self, error, dt):
        '''
        Update the control input based on error using PID formula.
        :param error: Current error
        :param dt: Time step
        :return: Control input
        '''
        self.integral += error * dt
        derivative = (error - self.previous_error) / dt
        self.previous_error = error

        return self.Kp * error + self.Ki * self.integral + self.Kd *
        ↪ derivative

# Rotational Dynamics
def rotational_dynamics(t, y, I, M):
    '''
    Rotational dynamics using Euler equations.
    :param t: Time
    :param y: State vector (angular velocities)
    :param I: Inertia tensor
    :param M: Moment vector
    :return: Derivative of state vector
    '''
    omega = y
    omega_dot = np.linalg.inv(I) @ (M - np.cross(omega, I @ omega))
    return omega_dot

# Example Usage
x_cg = 5.0  # Example center of gravity position
x_cp = 4.5  # Example center of pressure position
```

```python
# Static Stability Check
print("Statically Stable:", is_statically_stable(x_cg, x_cp))

# Dynamic Stability Check with example system matrix
A = np.array([[-0.5, 0.0], [0.1, -0.3]])
print("Dynamically Stable:", is_dynamically_stable(A))

# PID Control Simulation
def control_simulation():
    pid = PIDController(Kp=1.0, Ki=0.1, Kd=0.05)
    setpoint = 1.0
    dt = 0.1
    num_steps = 100
    control_input = []

    for step in range(num_steps):
        current_value = np.sin(step * dt)  # Simulated current
            ↪ value
        error = setpoint - current_value
        u = pid.update(error, dt)
        control_input.append(u)

    plt.plot(control_input)
    plt.title("Control Input Over Time")
    plt.xlabel("Time steps")
    plt.ylabel("Control Input")
    plt.show()

control_simulation()

# Rotational Dynamics Simulation
def rotational_simulation():
    I = np.diag([10.0, 10.0, 10.0])
    M = np.array([0.0, 1.0, 0.0])
    initial_omega = np.array([0.0, 0.0, 0.1])

    sol = solve_ivp(rotational_dynamics, [0, 10], initial_omega,
        ↪ args=(I, M), dense_output=True)
    t = np.linspace(0, 10, 500)
    omega = sol.sol(t)
    plt.plot(t, omega.T)
    plt.title("Angular Velocities Over Time")
    plt.xlabel("Time")
    plt.ylabel("Angular Velocities")
    plt.legend(["_x", "_y", "_z"])
    plt.show()

rotational_simulation()
```

This code defines various key functions and classes necessary for computations related to rocket stability and control:

- `is_statically_stable` determines if the rocket is statically stable by comparing the positions of the center of gravity and pressure.

- `is_dynamically_stable` checks dynamic stability using the eigenvalues of a system matrix to ensure they have negative real parts.

- `PIDController` is a class implementing a PID controller, which provides a control input based on proportional, integral, and derivative terms.

- `rotational_dynamics` uses Euler's equations to simulate the rotational motion of a rocket with given inertia and moment inputs.

The simulation examples (`control_simulation` and `rotational_simulation`) demonstrate how control input evolves over time using a PID controller and how angular velocities change using rotational dynamics.

Chapter 21

Quaternion Mathematics for Attitude Control

Introduction to Quaternions

Quaternions, first introduced by William Rowan Hamilton in 1843, serve as an extension of complex numbers and are particularly advantageous for three-dimensional rotational operations. They provide a robust representation of orientation, avoiding gimbal lock and offering smooth interpolations applicable in aerospace engineering.

A quaternion **q** is expressed as:

$$\mathbf{q} = q_w + q_x\mathbf{i} + q_y\mathbf{j} + q_z\mathbf{k} \qquad (21.1)$$

where q_w, q_x, q_y, and q_z are real numbers. For attitude control, quaternions convey rotation by means of a scalar q_w and a three-dimensional vector component (q_x, q_y, q_z).

Quaternion Algebra and Operations

The sum of two quaternions $\mathbf{q}_1$ and $\mathbf{q}_2$ is computed as:

$$\mathbf{q}_1+\mathbf{q}_2 = (q_{1w}+q_{2w})+(q_{1x}+q_{2x})\mathbf{i}+(q_{1y}+q_{2y})\mathbf{j}+(q_{1z}+q_{2z})\mathbf{k} \qquad (21.2)$$

Multiplication of two quaternions, fundamental for combining rotations, is given by:

$$\mathbf{q}_1 \cdot \mathbf{q}_2 = (q_{1w}q_{2w} - q_{1x}q_{2x} - q_{1y}q_{2y} - q_{1z}q_{2z}) \tag{21.3}$$
$$+ (q_{1w}q_{2x} + q_{1x}q_{2w} + q_{1y}q_{2z} - q_{1z}q_{2y})\mathbf{i} \tag{21.4}$$
$$+ (q_{1w}q_{2y} - q_{1x}q_{2z} + q_{1y}q_{2w} + q_{1z}q_{2x})\mathbf{j} \tag{21.5}$$
$$+ (q_{1w}q_{2z} + q_{1x}q_{2y} - q_{1y}q_{2x} + q_{1z}q_{2w})\mathbf{k} \tag{21.6}$$

Quaternion Representation of Rotations

For a vector $\mathbf{v} = (v_x, v_y, v_z)$, its rotation by a quaternion $\mathbf{q} = (q_w, q_x, q_y, q_z)$ is performed through:

$$\mathbf{v}' = \mathbf{q} \cdot \mathbf{v} \cdot \mathbf{q}^* \tag{21.7}$$

where $\mathbf{q}^* = (q_w, -q_x, -q_y, -q_z)$ is the conjugate of the quaternion $\mathbf{q}$.

Quaternion Kinematics

The quaternion derivative for representing attitude rate or angular velocity $\boldsymbol{\omega}$ is:

$$\dot{\mathbf{q}} = \frac{1}{2}\mathbf{q} \cdot \boldsymbol{\omega} \tag{21.8}$$

Quaternions simplify the computational structure in feedback systems, contributing to the robustness of aerospace application, such as stabilizing a spacecraft's orientation in the absence of external torques.

Application in Control Systems

In control systems, quaternions aid in the development of control laws essential for tracking desired orientations or maneuvers. The error quaternion $\mathbf{q}_e$, representing the deviation from a desired orientation $\mathbf{q}_d$, is calculated as:

$$\mathbf{q}_e = \mathbf{q}_d \cdot \mathbf{q}^{-1} \tag{21.9}$$

This error quaternion underpins control strategies linked with Proportional-Derivative (PD) controllers. The control torque $\mathbf{M}_c$ can be computed via:

$$\mathbf{M}_c = -K_p \mathbf{q}_e - K_d \boldsymbol{\omega} \qquad (21.10)$$

where K_p and K_d are proportional and derivative gain matrices.

Integration in Guidance Systems

Quaternion-based algorithms are pivotal for the integration in guidance systems, streamlining computation of angular positions and velocities necessary for rocket or spacecraft navigation. Guidance laws leverage continuous quaternion updating, maintaining the vehicle's alignment with mission trajectories.

The quaternion-based rotation matrix $\mathbf{R}$ for transforming vectors between coordinate frames is defined as:

$$\mathbf{R}(\mathbf{q}) = \begin{bmatrix} 1 - 2(q_y^2 + q_z^2) & 2(q_x q_y - q_w q_z) & 2(q_x q_z + q_w q_y) \\ 2(q_x q_y + q_w q_z) & 1 - 2(q_x^2 + q_z^2) & 2(q_y q_z - q_w q_x) \\ 2(q_x q_z - q_w q_y) & 2(q_y q_z + q_w q_x) & 1 - 2(q_x^2 + q_y^2) \end{bmatrix}$$
$$(21.11)$$

Python Code Snippet

Below is a Python code snippet that encompasses the core computational elements of quaternion mathematics and its application in attitude control systems, including operations such as addition, multiplication, rotation, kinematics, and control torque computation.

```python
import numpy as np

class Quaternion:
    def __init__(self, w, x, y, z):
        self.w = w
        self.x = x
        self.y = y
        self.z = z

    def __add__(self, other):
        ''' Addition of two quaternions '''
        return Quaternion(self.w + other.w,
```

```python
                        self.x + other.x,
                        self.y + other.y,
                        self.z + other.z)

    def __mul__(self, other):
        ''' Multiplication of two quaternions '''
        w = self.w * other.w - self.x * other.x - self.y * other.y -
            self.z * other.z
        x = self.w * other.x + self.x * other.w + self.y * other.z -
            self.z * other.y
        y = self.w * other.y - self.x * other.z + self.y * other.w +
            self.z * other.x
        z = self.w * other.z + self.x * other.y - self.y * other.x +
            self.z * other.w
        return Quaternion(w, x, y, z)

    def conjugate(self):
        ''' Conjugate of the quaternion '''
        return Quaternion(self.w, -self.x, -self.y, -self.z)

    def rotate_vector(self, v):
        ''' Rotate a vector using quaternion '''
        q_vec = Quaternion(0, *v)
        rotated_vec = self * q_vec * self.conjugate()
        return rotated_vec.x, rotated_vec.y, rotated_vec.z

    @staticmethod
    def from_angular_velocity(w, dt):
        ''' Quaternion derivative for angular velocity '''
        q_dot = Quaternion(0, *w) * dt * 0.5
        return q_dot

def control_torque(q_error, omega, Kp, Kd):
    ''' Calculate control torque '''
    return -Kp * np.array([q_error.x, q_error.y, q_error.z]) - Kd *
        omega

# Example usage
q1 = Quaternion(0.7071, 0.7071, 0, 0)
q2 = Quaternion(0.7071, 0, 0.7071, 0)

# Quaternion operations
q_sum = q1 + q2
q_product = q1 * q2

# Rotation example
v = np.array([1, 0, 0])
v_rotated = q1.rotate_vector(v)

# Quaternion derivative from angular velocity
omega = np.array([0.01, 0.01, 0.02])
dt = 1.0
q_dot = Quaternion.from_angular_velocity(omega, dt)
```

```
# Control torque calculation
q_error = Quaternion(0.0, 0.1, 0.1, 0.1)
Kp = np.eye(3)  # Proportional gain
Kd = np.eye(3)  # Derivative gain
torque = control_torque(q_error, omega, Kp, Kd)

print("Quaternion Sum:", q_sum.w, q_sum.x, q_sum.y, q_sum.z)
print("Quaternion Product:", q_product.w, q_product.x, q_product.y,
      q_product.z)
print("Rotated Vector:", v_rotated)
print("Quaternion Derivative:", q_dot.w, q_dot.x, q_dot.y, q_dot.z)
print("Control Torque:", torque)
```

This code defines key functionalities related to quaternion-based attitude control systems:

- **Quaternion** class implements essential quaternion operations such as addition and multiplication, including methods for rotation and conjugate calculations.

- **rotate_vector** method rotates a vector based on quaternion orientation, representing real-world application in aerospace dynamics.

- **from_angular_velocity** provides quaternion derivatives for given angular velocities, essential for quaternion kinematics in motion planning.

- **control_torque** computes the control torque needed for stabilizing deviations from desired orientations, leveraging quaternion error states.

The code block demonstrates quaternion operations and computations using example data for practical insight into their application in control systems.

Chapter 22

Gyroscopic Effects and Calculations

Fundamentals of Gyroscopic Motion

Gyroscopic motion arises from the conservation of angular momentum and is critical in maintaining the stability of aerospace vehicles, particularly rockets. The angular momentum **L** of a gyroscope is given by:

$$\mathbf{L} = \mathbf{I} \cdot \boldsymbol{\omega}$$

where **I** is the moment of inertia tensor, and $\boldsymbol{\omega}$ is the angular velocity vector. In control systems, the gyroscopic principles are leveraged to resist changes in orientation, providing stability to the vehicle's flight trajectory.

Mathematical Model of Gyroscopic Stability

In the context of rocket flight, gyroscopic stability can be mathematically represented by the torque $\boldsymbol{\tau}$ experienced due to angular momentum variations, which is expressed as:

$$\boldsymbol{\tau} = \frac{d\mathbf{L}}{dt} = \mathbf{I} \cdot \frac{d\boldsymbol{\omega}}{dt} + \boldsymbol{\omega} \times (\mathbf{I} \cdot \boldsymbol{\omega})$$

The above expression indicates how the change in angular velocity and any existing angular velocity affect the torque on the gyroscopic system, critical for maintaining desired flight stability.

Gyroscopic Precession and Nutation

Gyroscopic precession occurs when an external torque is applied, causing the axis of rotation to move around in a circular path. The precession angular velocity Ω of a system subjected to an external torque τ is given by:

$$\Omega = \frac{\tau}{L}$$

Nutation, on the other hand, refers to the oscillatory motion superimposed on the precession. It is characterized by its frequency and amplitude, which are dependent on the initial conditions of the system and the applied forces.

Application of Gyroscopic Principles in Flight Control

In aerospace applications, particularly rocket guidance systems, gyroscopic sensors termed 'gyroscopes' are implemented to measure and control changes in orientation. Using the fundamental equations of gyroscopic motion, control algorithms calculate the necessary adjustments to the propulsion or control surfaces to maintain stability.

The control strategy typically includes feedback loops that incorporate gyroscopic measurements $\mathbf{g}$ into the control laws. The control torque $\mathbf{T}_c$ is computed as:

$$\mathbf{T}_c = -K_g(\mathbf{g} - \mathbf{g}_d) - K_v \dot{\mathbf{g}}$$

where K_g and K_v are gain matrices, $\mathbf{g}$ represents the current gyroscopic readings, and $\mathbf{g}_d$ is the desired gyroscopic vector. This feedback mechanism is essential for suppressing undesired rotational motion and achieving intended flight dynamics.

Python Code Snippet

Below is a Python code snippet that encompasses the core computational elements of gyroscopic effects calculations, including the evaluation of angular momentum, torque determination, precession, and application in flight control:

```python
import numpy as np

def angular_momentum(I, omega):
    '''
    Calculate angular momentum.
    :param I: Moment of inertia tensor.
    :param omega: Angular velocity vector.
    :return: Angular momentum vector.
    '''
    return np.dot(I, omega)

def gyroscopic_torque(dL_dt, I, omega):
    '''
    Calculate the gyroscopic torque.
    :param dL_dt: Derivative of angular momentum.
    :param I: Moment of inertia tensor.
    :param omega: Angular velocity vector.
    :return: Torque vector.
    '''
    return dL_dt + np.cross(omega, np.dot(I, omega))

def precession_angular_velocity(torque, L):
    '''
    Calculate precession angular velocity.
    :param torque: External torque vector.
    :param L: Angular momentum vector.
    :return: Precession angular velocity vector.
    '''
    return torque / np.linalg.norm(L)

def flight_control_torque(g, g_d, K_g, K_v, g_dot):
    '''
    Calculate control torque for flight stability.
    :param g: Current gyroscopic measurements.
    :param g_d: Desired gyroscopic vector.
    :param K_g: Gain matrix for gyroscopic measurements.
    :param K_v: Gain matrix for derivative of gyroscopic
        measurements.
    :param g_dot: Derivative of gyroscopic measurements.
    :return: Control torque.
    '''
    return -np.dot(K_g, (g - g_d)) - np.dot(K_v, g_dot)

# Example of gyroscope parameters and control calculations
```

```
I = np.array([[1, 0, 0], [0, 1, 0], [0, 0, 1]])  # Identity matrix
↪ as a simple inertia tensor
omega = np.array([0.1, 0.1, 0.1])  # Angular velocity vector

L = angular_momentum(I, omega)
dL_dt = np.array([0.01, 0.01, 0.01])  # Hypothetical change in
↪ angular momentum
torque = gyroscopic_torque(dL_dt, I, omega)
precession_velocity = precession_angular_velocity(torque, L)

K_g = np.eye(3)  # Gain matrix for proportional control
K_v = np.eye(3) * 0.5  # Gain matrix for derivative control
g = np.array([0.12, 0.12, 0.12])
g_d = np.array([0, 0, 0])  # Desired gyroscopic vector
g_dot = np.array([0.01, -0.02, 0.00])  # Rate of change

control_torque = flight_control_torque(g, g_d, K_g, K_v, g_dot)

print("Angular Momentum:", L)
print("Gyroscopic Torque:", torque)
print("Precession Angular Velocity:", precession_velocity)
print("Control Torque:", control_torque)
```

This code defines several key functions necessary for the implementation and evaluation of gyroscopic effects:

- angular_momentum function computes the angular momentum given the inertia tensor and angular velocity.

- gyroscopic_torque calculates the torque based on changing angular momentum and current conditions.

- precession_angular_velocity determines the precession rate under an applied torque.

- flight_control_torque computes the control torque essential for maintaining stability in flight control systems.

The final block of code demonstrates the use of these functions with example data to achieve desired stability and control in aerospace applications.

Chapter 23

Calculating Structural Loads

Introduction to Structural Load Analysis

The assessment of structural loads in aerospace engineering is critical to ensure the integrity and safety of rocket designs. Structural loads are the forces, moments, and thermal influences acting upon a rocket, which must be evaluated meticulously during the design phase. The accurate prediction and analysis of these loads involve various complex methodologies, pivotal for planning and executing successful launches.

Stress and Strain Fundamentals

Understanding the stress and strain behavior of materials under load is fundamental. Stress (σ) is defined as the internal force per unit area within materials, expressed in terms of the applied load F and the cross-sectional area A:

$$\sigma = \frac{F}{A}$$

Strain (ϵ) characterizes the deformation experienced by a material, calculated as the ratio of the change in length ΔL to the original length L_0:

$$\epsilon = \frac{\Delta L}{L_0}$$

Hooke's Law describes the linear relationship between stress and strain for elastic materials, characterized by the modulus of elasticity E:

$$\sigma = E \cdot \epsilon$$

Load Types and Distribution

Rockets experience various types of loads during flight, including axial, shear, torsional, and bending loads. Each of these loads exerts distinctive stress distributions across the rocket structure. The axial load (P) leads to compressive or tensile stress, crucial during stages like launch and landing. In contrast, shear loads result from forces parallel to the surface, producing shear stress (τ):

$$\tau = \frac{V}{A}$$

where V is the shear force.

Torsional loads arise from twisting actions, creating a torque T. The associated shear stress τ_t in a material with radius r relates to the polar moment of inertia J as follows:

$$\tau_t = \frac{T \cdot r}{J}$$

Dynamic Loading Conditions

Rockets are subject to dynamic loads due to changing flight conditions, such as vibrations, buffeting, and rapid accelerations. These loads are inherently time-dependent and necessitate the use of dynamic analysis techniques. The application of Newton's Second Law to the dynamic load analysis results in the following equation of motion:

$$m \cdot \frac{d^2 x}{dt^2} = \sum F(t)$$

where m is the mass of the structure and x represents displacement.

Finite Element Analysis for Complex Structures

Finite Element Analysis (FEA) serves as an indispensable tool for assessing structural integrity under complex loading conditions. Aerospace structures are discretized into finite elements, allowing the computational evaluation of stress and strain. Utilizing the stiffness matrix **K**, the equilibrium among nodes **u** is expressed as:

$$\mathbf{K} \cdot \mathbf{u} = \mathbf{F}$$

where **F** signifies the applied nodal forces. FEA aids in accurately modeling stress concentrations and deformation patterns essential during design evaluations.

Thermal Loads and Material Responses

The propulsion process and external environmental conditions subject rockets to thermal loads, causing expansion or contraction. Thermal stress σ_t is given by:

$$\sigma_t = E \cdot \alpha \cdot \Delta T$$

where α represents the coefficient of thermal expansion, and ΔT is the temperature change. Accurate thermomechanical analysis is necessary to prevent structural failures due to temperature variations.

Load Factors and Safety Margins

In aerospace applications, considering load factors and implementing safety margins is crucial to counteract uncertainties. The factor of safety n is defined as the ratio of the allowable stress $\sigma_{\text{allowable}}$ to the actual stress σ_{actual}:

$$n = \frac{\sigma_{\text{allowable}}}{\sigma_{\text{actual}}}$$

Designs are iteratively tested to ensure that safety margins are adequate for anticipated load conditions.

Python Code Snippet

Below is a Python code snippet that encompasses the core mathematical and computational components addressing structural load calculations including stress, strain, dynamic loading, and the use of finite element analysis in aerospace applications.

```python
import numpy as np

def stress(force, area):
    '''
    Calculate the stress applied to a material.
    :param force: The force applied (N).
    :param area: The cross-sectional area (m^2).
    :return: Stress value (Pa).
    '''
    return force / area

def strain(delta_length, original_length):
    '''
    Calculate the strain experienced by a material.
    :param delta_length: Change in length (m).
    :param original_length: Original length (m).
    :return: Strain value.
    '''
    return delta_length / original_length

def hookes_law(stress, modulus_of_elasticity):
    '''
    Calculate strain using Hooke's Law.
    :param stress: Stress value (Pa).
    :param modulus_of_elasticity: Modulus of elasticity (Pa).
    :return: Strain value.
    '''
    return stress / modulus_of_elasticity

def shear_stress(shear_force, area):
    '''
    Calculate the shear stress.
    :param shear_force: The shear force applied (N).
    :param area: The cross-sectional area (m^2).
    :return: Shear stress value (Pa).
    '''
    return shear_force / area

def torsional_shear_stress(torque, radius, polar_moment_of_inertia):
    '''
    Calculate the torsional shear stress.
    :param torque: The torque applied (Nm).
    :param radius: The radius (m).
```

```
    :param polar_moment_of_inertia: The polar moment of inertia
    ↪    (m^4).
    :return: Torsional shear stress value (Pa).
    '''
    return torque * radius / polar_moment_of_inertia

def dynamic_load_analysis(mass, displacement, force_time_function):
    '''
    Analyze dynamic load using Newton's Second Law.
    :param mass: Mass of the structure (kg).
    :param displacement: Displacement (m).
    :param force_time_function: Function returning force at given
    ↪    time t (N).
    :return: Acceleration value (m/s^2).
    '''
    total_force = force_time_function()
    return total_force / mass

def finite_element_analysis(stiffness_matrix, nodal_forces):
    '''
    Perform a finite element analysis calculation.
    :param stiffness_matrix: Stiffness matrix (2D array).
    :param nodal_forces: Nodal forces (array).
    :return: Nodal displacements (array).
    '''
    return np.linalg.solve(stiffness_matrix, nodal_forces)

def thermal_stress(modulus_of_elasticity,
↪   coefficient_of_thermal_expansion, temperature_change):
    '''
    Calculate thermal stress.
    :param modulus_of_elasticity: Modulus of elasticity (Pa).
    :param coefficient_of_thermal_expansion: Coefficient of thermal
    ↪    expansion (/°C).
    :param temperature_change: Change in temperature (°C).
    :return: Thermal stress value (Pa).
    '''
    return modulus_of_elasticity * coefficient_of_thermal_expansion
    ↪    * temperature_change

def factor_of_safety(allowable_stress, actual_stress):
    '''
    Calculate the factor of safety.
    :param allowable_stress: Allowable stress (Pa).
    :param actual_stress: Actual stress (Pa).
    :return: Factor of safety.
    '''
    return allowable_stress / actual_stress

# Example values for demonstration
F = 1000    # N
A = 0.01    # m^2
E = 200e9   # Pa
```

```
alpha = 12e-6     # /°C
delta_T = 50      # °C
torque = 500      # Nm
radius = 0.05     # m
J = 1e-6          # m^4
mass = 100        # kg
displacement = 0.002   # m

# Placeholder function returning constant force
def force_time_function():
    return 5000   # N

# Finite element example
K = np.array([[10, -2], [-2, 5]])  # Stiffness matrix (toy example)
F_nodal = np.array([100, 200])     # Nodal forces

# Calculate stress and strain
sigma = stress(F, A)
epsilon = strain(0.001, 0.1)
sigma_t = thermal_stress(E, alpha, delta_T)

# Calculate dynamics
acceleration = dynamic_load_analysis(mass, displacement,
↪    force_time_function)

# Calculate FEA
nodal_displacements = finite_element_analysis(K, F_nodal)

# Safety factor example
safety_factor = factor_of_safety(300e6, sigma)

print("Stress:", sigma)
print("Strain:", epsilon)
print("Thermal Stress:", sigma_t)
print("Nodal Displacements:", nodal_displacements)
print("Factor of Safety:", safety_factor)
```

This code defines several key functions necessary for structural load analysis in aerospace engineering:

- `stress` and `shear_stress` functions compute stress values based on applied forces and areas.

- `strain` calculates material deformation relative to its original length.

- `hookes_law` applies Hooke's Law to find strain for elastic materials.

- `dynamic_load_analysis` analyzes dynamic loads through time-dependent force evaluations.

- `finite_element_analysis` performs basic FEA, solving nodal displacements.

- `thermal_stress` evaluates stress due to temperature variations.

- `factor_of_safety` calculates safety margins within structural components.

The example outputs demonstrate how these computations are conducted using representative inputs.

Chapter 24

Heat Shield Calculations

Heat Transfer Mechanisms

Aerospace vehicles, particularly rockets, endure extreme thermal environments that demand effective thermal protection systems (TPS). Heat shield calculations necessitate a deep understanding of heat transfer mechanisms, primarily conduction, convection, and radiation. The conduction of heat through solid materials is described by Fourier's Law, given by:

$$q = -k\nabla T$$

where q represents the heat flux, k is the thermal conductivity, and ∇T is the temperature gradient.

Convective heat transfer between the rocket surface and the surrounding fluid is expressed in terms of Newton's Law of Cooling:

$$q = h(T_s - T_\infty)$$

where h is the convective heat transfer coefficient, T_s is the surface temperature, and T_∞ is the fluid temperature away from the surface.

Radiative heat transfer follows the Stefan-Boltzmann Law, articulated as:

$$q = \epsilon\sigma T^4$$

where ϵ is the emissivity of the surface and σ is the Stefan-Boltzmann constant.

Thermal Response Models

The efficacy of heat shield materials is often evaluated through thermal response models. One-dimensional transient heat conduction is governed by the heat equation:

$$\frac{\partial T}{\partial t} = \alpha \nabla^2 T$$

where α denotes the thermal diffusivity of the material ($\alpha = \frac{k}{\rho c_p}$, ρ and c_p being the density and specific heat capacity, respectively).

For ablative materials, energy conservation during recession involves ablative rate equations, such as:

$$\dot{m} = \frac{q - \dot{m} \cdot L}{H_c}$$

where $\dot{m}$ is the mass loss rate, L the latent heat of ablation, and H_c the convective heat transfer input.

Material Selection and Performance

Material selection for heat shields involves a balance between thermal, mechanical, and physical properties. The efficiency factor η of a thermal protection material can be estimated by:

$$\eta = \frac{c_p \cdot k \cdot \rho \cdot \Delta T}{E_c}$$

where E_c is the total energy encounter throughout the mission profile. The optimal design minimizes the peak heat flux $q_{\max}$ on the vehicle's surface, crucial for maintaining structural integrity.

Thermo-mechanical Stress Analysis

Evaluation of the thermo-mechanical stresses is essential given the significant temperature gradients induced during high-speed atmospheric entry. Such stress can be quantified using the thermal stress equation:

$$\sigma_t = E \cdot \alpha \cdot \Delta T$$

This equation assumes a linear thermal expansion coefficient α and moduli of elasticity E that are thermal-conditions dependent.

Analytical solutions often implement temperature distribution calculations through the use of finite difference methods, specifically for complex geometries:

$$A \frac{d}{dx}\left(k \frac{dT}{dx}\right) = \dot{q}$$

where A is the cross-sectional area, and $\dot{q}$ is the internal heat generation rate.

Validation against Empirical Data

Thermal protection systems' design and analysis are iterative, involving experimental validation against empirical datasets and wind tunnel test results. The iterative refinement processes frequently employ computational fluid dynamics (CFD) to forecast external aero-thermal conditions encountered during different mission stages and validate convective heat transfer coefficients.

Empirical relationships, applicable in stagnation regions, include the Fay-Riddell equation. It predicts the peak coefficient of heat transfer:

$$h_s = 0.5 \cdot (\rho_\infty \cdot v_\infty \cdot c_p \cdot \text{Pr}^{-0.6})^{0.5}$$

Here, ρ_∞ and v_∞ denote the free stream density and velocity, respectively, and Pr is the Prandtl number.

Future Considerations

As aerospace projects push the limits of atmospheric re-entry speeds, the development of advanced thermal protection materials, such as metallic and carbon-carbon composites, and innovative design concepts, is critical. Incorporating these advanced materials and models allows for the improved prediction and management of temperature profiles across mission-specific parameters.

Python Code Snippet

Below is a Python code snippet that encompasses the core computational elements of heat shield calculations including heat transfer mechanisms, thermal response models, material performance evaluation, thermo-mechanical stress analysis, and simulation readiness.

```python
import numpy as np

def calculate_heat_flux_conduction(k, grad_T):
    '''
    Calculate the heat flux based on conduction.
    :param k: Thermal conductivity.
    :param grad_T: Temperature gradient.
    :return: Heat flux by conduction.
    '''
    return -k * grad_T

def calculate_heat_flux_convection(h, T_s, T_inf):
    '''
    Calculate the heat flux based on convection.
    :param h: Convective heat transfer coefficient.
    :param T_s: Surface temperature.
    :param T_inf: Fluid temperature away from the surface.
    :return: Heat flux by convection.
    '''
    return h * (T_s - T_inf)

def calculate_heat_flux_radiation(epsilon, sigma, T):
    '''
    Calculate the heat flux based on radiation.
    :param epsilon: Emissivity of the surface.
    :param sigma: Stefan-Boltzmann constant.
    :param T: Temperature of the surface.
    :return: Heat flux by radiation.
    '''
    return epsilon * sigma * T**4

def thermal_response(alpha, T, delta_t, grad_T2):
    '''
    Model transient heat conduction response.
    :param alpha: Thermal diffusivity.
    :param T: Temperature array.
    :param delta_t: Time step.
    :param grad_T2: Second derivative of temperature (Laplacian).
    :return: Updated temperature.
    '''
    return T + alpha * grad_T2 * delta_t

def calculate_ablation_rate(q, dot_m, L, H_c):
    '''
```

```python
    Calculate mass loss rate in ablative materials.
    :param q: Convective heat transfer input.
    :param dot_m: Initial mass loss rate.
    :param L: Latent heat of ablation.
    :param H_c: Convective heat absorbed.
    :return: Mass loss rate.
    '''
    return (q - dot_m * L) / H_c

def efficiency_factor(cp, k, rho, delta_T, E_c):
    '''
    Calculate efficiency factor of thermal protection.
    :param cp: Specific heat capacity.
    :param k: Thermal conductivity.
    :param rho: Density.
    :param delta_T: Temperature difference.
    :param E_c: Total energy.
    :return: Efficiency factor.
    '''
    return (cp * k * rho * delta_T) / E_c

def thermal_stress(E, alpha, delta_T):
    '''
    Calculate thermal stress.
    :param E: Modulus of elasticity.
    :param alpha: Linear thermal expansion coefficient.
    :param delta_T: Temperature change.
    :return: Thermal stress.
    '''
    return E * alpha * delta_T

def fay_riddell_equation(rho_inf, v_inf, cp, Pr):
    '''
    Calculate the peak coefficient of heat transfer in stagnation.
    :param rho_inf: Free stream density.
    :param v_inf: Free stream velocity.
    :param cp: Specific heat at constant pressure.
    :param Pr: Prandtl number.
    :return: Stagnation heat transfer coefficient.
    '''
    return 0.5 * (rho_inf * v_inf * cp * Pr**-0.6)**0.5

# Constants
sigma = 5.67e-8  # Stefan-Boltzmann constant
E = 200e9  # Modulus of elasticity (Example value)
alpha_thermal = 1e-5  # Thermal expansion coefficient

# Example inputs
k = 0.1  # Thermal conductivity
grad_T = 100  # Temperature gradient
h = 10  # Convective heat transfer coefficient
T_s = 1000  # Surface temperature
T_inf = 200  # Fluid temperature away
```

```python
epsilon = 0.9  # Emissivity
T_rad = 800  # Temperature for radiation
alpha = 0.01  # Thermal diffusivity
T_current = 300  # Current temperature
delta_t = 0.1  # Time step
grad_T2 = 10  # Second derivative of temperature
dot_m_initial = 0.01  # Initial mass loss rate
L = 2.5e6  # Latent heat of ablation
H_c = 0.02  # Convective heat absorbed
cp = 1000  # Specific heat capacity
rho = 2700  # Density
delta_T = 100  # Temperature change
E_c = 10e4  # Total energy encounter
rho_inf = 1.2  # Free stream density
v_inf = 300  # Free stream velocity
Pr = 0.72  # Prandtl number

# Outputs for demonstration
q_conduction = calculate_heat_flux_conduction(k, grad_T)
q_convection = calculate_heat_flux_convection(h, T_s, T_inf)
q_radiation = calculate_heat_flux_radiation(epsilon, sigma, T_rad)
T_updated = thermal_response(alpha, T_current, delta_t, grad_T2)
ablation_rate = calculate_ablation_rate(q_convection, dot_m_initial,
    L, H_c)
efficiency = efficiency_factor(cp, k, rho, delta_T, E_c)
stress = thermal_stress(E, alpha_thermal, delta_T)
h_s = fay_riddell_equation(rho_inf, v_inf, cp, Pr)

print("Conduction Heat Flux:", q_conduction)
print("Convection Heat Flux:", q_convection)
print("Radiation Heat Flux:", q_radiation)
print("Updated Temperature:", T_updated)
print("Ablation Rate:", ablation_rate)
print("Efficiency Factor:", efficiency)
print("Thermal Stress:", stress)
print("Stagnation Heat Transfer Coefficient:", h_s)
```

This code defines several key functions necessary for the implementation and evaluation of heat shield calculations:

- calculate_heat_flux_conduction function computes the heat flux due to conduction.

- calculate_heat_flux_convection calculates the heat flux based on convection parameters.

- calculate_heat_flux_radiation evaluates the heat flux from radiation.

- thermal_response models the transient thermal response of materials.

- `calculate_ablation_rate` determines the mass loss rate from ablation factors.

- `efficiency_factor` calculates the efficiency factor for thermal protection materials.

- `thermal_stress` assesses the thermal stresses induced by temperature gradients.

- `fay_riddell_equation` applies to heat transfer prediction in stagnation regions.

The final block of code provides examples of computing these elements using given constant and variable inputs.

Chapter 25

Thermal Control Systems for Spacecraft

Heat Transfer in Spacecraft

The thermal control of spacecraft involves a complex interplay of heat rejection, absorption, and regulation mechanisms to maintain operational temperature ranges. Given the vacuum of space, primary modes of heat transfer are conduction and radiation, with negligible convection. The conduction through spacecraft structural elements can be represented by Fourier's Law:

$$q = -k\nabla T \tag{25.1}$$

where k denotes the thermal conductivity, and ∇T represents the temperature gradient.

Radiative heat transfer between spacecraft surfaces follows the Stefan-Boltzmann Law:

$$q = \epsilon \sigma A (T^4 - T_{\text{env}}^4) \tag{25.2}$$

where ϵ is the emissivity, σ the Stefan-Boltzmann constant, A the surface area, and T_{env} the temperature of the surrounding environment.

Thermal Control Mechanisms

Spacecraft employ various thermal control elements such as radiators, heat pipes, and thermal blankets. The effectiveness of a radiator can be assessed through its heat rejection capacity:

$$Q_{\text{rad}} = \epsilon \sigma A_{\text{rad}} T_{\text{rad}}^4 \qquad (25.3)$$

Calculating the thermal resistance (R_{th}) of composite materials is crucial for evaluating heat conduction paths:

$$R_{\text{th}} = \frac{L}{kA} \qquad (25.4)$$

where L is the thickness of the material, and A the cross-sectional area.

Active Thermal Regulation

Active thermal control systems utilize mechanisms like heaters and thermoelectric coolers. The power requirement for resistive heaters is given by:

$$P_{\text{heater}} = I^2 R \qquad (25.5)$$

where I is the current and R the resistance of the heater element.

For thermoelectric coolers, the heat pumped can be described by:

$$Q_{\text{TEC}} = \alpha I T - \frac{1}{2} I^2 R - K \Delta T \qquad (25.6)$$

where α is the Seebeck coefficient, T the absolute temperature, K the thermal conductance, and ΔT the temperature difference across the cooler.

Passive Thermal Control Strategies

Passive thermal control relies on minimizing and controlling heat fluctuations. Multilayer insulation (MLI) effectiveness is calculated by reducing radiative exchange between spacecraft surfaces:

$$q_{\text{MLI}} = \frac{\sigma \epsilon_{\text{eff}} A_{\text{MLI}} (T_1^4 - T_2^4)}{N + 1} \qquad (25.7)$$

where N is the number of layers, and ϵ_{eff} is the effective emissivity.

Thermal Analysis Techniques

Thermal analysis of spacecraft systems necessitates solving complex systems of equations that define heat transfer. Finite difference methods are often employed for one-dimensional transient heat conduction:

$$\frac{\partial T}{\partial t} = \alpha \frac{\partial^2 T}{\partial x^2} \qquad (25.8)$$

where α is the thermal diffusivity.
endlatex

Python Code Snippet

Below is a Python code snippet that illustrates the core computational elements involved in the thermal control of spacecraft, including calculations for heat transfer, thermal resistance, power requirements for heaters, and radiative heat transfer efficiency.

```
import numpy as np

def fouriers_conduction(k, grad_T):
    '''
    Calculate heat transfer using Fourier's Law.
    :param k: Thermal conductivity.
    :param grad_T: Temperature gradient.
    :return: Heat transfer due to conduction.
    '''
    return -k * grad_T

def stefan_boltzmann_radiation(epsilon, sigma, A, T, T_env):
    '''
    Calculate radiative heat transfer using the Stefan-Boltzmann
    ↪ Law.
    :param epsilon: Emissivity.
    :param sigma: Stefan-Boltzmann constant.
    :param A: Surface area.
    :param T: Surface temperature.
```

```python
    :param T_env: Environment temperature.
    :return: Radiative heat transfer.
    '''
    return epsilon * sigma * A * (T**4 - T_env**4)

def thermal_resistance(L, k, A):
    '''
    Calculate thermal resistance of a material.
    :param L: Thickness of the material.
    :param k: Thermal conductivity.
    :param A: Cross-sectional area.
    :return: Thermal resistance.
    '''
    return L / (k * A)

def heater_power(I, R):
    '''
    Calculate power requirement for a resistive heater.
    :param I: Current through the heater.
    :param R: Resistance of the heater.
    :return: Power required by the heater.
    '''
    return I**2 * R

def tec_heat_pumped(alpha, I, T, R, K, delta_T):
    '''
    Calculate the heat pumped by a thermoelectric cooler.
    :param alpha: Seebeck coefficient.
    :param I: Current.
    :param T: Absolute temperature.
    :param R: Electrical resistance of the TEC.
    :param K: Thermal conductance.
    :param delta_T: Temperature difference.
    :return: Heat pumped by the TEC.
    '''
    return alpha * I * T - 0.5 * I**2 * R - K * delta_T

def mli_heat_transfer(sigma, epsilon_eff, A, T1, T2, N):
    '''
    Calculate heat transfer through multilayer insulation.
    :param sigma: Stefan-Boltzmann constant.
    :param epsilon_eff: Effective emissivity.
    :param A: Surface area of MLI.
    :param T1: Temperature of the first surface.
    :param T2: Temperature of the second surface.
    :param N: Number of insulation layers.
    :return: Heat transfer through MLI.
    '''
    return (sigma * epsilon_eff * A * (T1**4 - T2**4)) / (N + 1)

def transient_heat_conduction(alpha, delta_x, delta_t, T_prev,
    ↪ T_curr):
    '''
```

```
    Solve 1D transient heat conduction using finite difference
↪   method.
    :param alpha: Thermal diffusivity.
    :param delta_x: Spatial step size.
    :param delta_t: Time step size.
    :param T_prev: Temperature at previous time step.
    :param T_curr: Current temperature array.
    :return: Updated temperature array.
    '''
    return T_curr + (alpha * delta_t / delta_x**2) *
↪   (np.roll(T_curr, 1) - 2 * T_curr + np.roll(T_curr, -1))

# Demonstration with example data
k = 205    # Thermal conductivity in W/(m K)
grad_T = np.array([0.01, 0.02, 0.01])    # Temperature gradient in K/m
epsilon = 0.8
sigma = 5.67e-8    # Stefan-Boltzmann constant in W/(m^2 K^4)
A = 1.0    # Surface area in m^2
T = 300    # Temperature in K
T_env = 77    # Environment temperature in K
L = 0.005    # Thickness in meters
I = 2    # Current in A
R = 10    # Resistance in ohms
alpha = 0.02    # Seebeck coefficient in V/K
K = 0.1    # Thermal conductance
delta_T = 20    # Temperature difference in K
N = 10    # Number of MLI layers
alpha_diffusivity = 1.4e-5    # m^2/s
delta_x = 0.01    # Spatial step size in m
delta_t = 0.1    # Time step size in seconds
T_prev = np.zeros(10)    # Previous time step temperatures
T_curr = np.ones(10) * 300    # Current time step temperatures

conduction_heat = fouriers_conduction(k, grad_T)
radiation_heat = stefan_boltzmann_radiation(epsilon, sigma, A, T,
↪   T_env)
thermal_res = thermal_resistance(L, k, A)
heater_p = heater_power(I, R)
heat_pumped = tec_heat_pumped(alpha, I, T, R, K, delta_T)
mli_heat = mli_heat_transfer(sigma, epsilon, A, T, T_env, N)
updated_temps = transient_heat_conduction(alpha_diffusivity,
↪   delta_x, delta_t, T_prev, T_curr)

print("Conduction Heat:", conduction_heat)
print("Radiation Heat:", radiation_heat)
print("Thermal Resistance:", thermal_res)
print("Heater Power:", heater_p)
print("Heat Pumped by TEC:", heat_pumped)
print("MLI Heat Transfer:", mli_heat)
print("Updated Temperatures:", updated_temps)
```

This code defines the essential functions and calculations used in spacecraft thermal management systems:

- `fouriers_conduction` computes heat transfer due to conduction using Fourier's law.

- `stefan_boltzmann_radiation` calculates radiative heat transfer as per the Stefan-Boltzmann law.

- `thermal_resistance` evaluates thermal resistance in materials.

- `heater_power` determines power consumption for electrical heaters.

- `tec_heat_pumped` computes heat management by thermoelectric coolers.

- `mli_heat_transfer` evaluates the effectiveness of multilayer insulation.

- `transient_heat_conduction` uses finite difference method for transient heat conduction simulations.

The demonstration code runs these computations using example data, providing insights into spacecraft thermal control scenarios.

Chapter 26

Escape Velocity and Gravity Assists

Escape Velocity Concept

Escape velocity is a fundamental aspect in the study of celestial mechanics, pivotal for understanding how an object can break free from a planet's gravitational influence. The concept relies on equating kinetic and gravitational potential energies. For an object of mass m attempting to escape a celestial body of mass M and radius R, the escape velocity v_e is derived as follows:

$$v_e = \sqrt{2\frac{GM}{R}} \tag{26.1}$$

where G denotes the gravitational constant. The escape velocity calculation assumes no atmospheric drag and a purely radial trajectory, conditions that simplify the model to first-order approximations.

Gravitational Potential Energy and Kinetic Energy Balance

The escape velocity necessitates that the sum of the kinetic energy $\left(\frac{1}{2}mv^2\right)$ and gravitational potential energy $\left(-\frac{GMm}{r}\right)$ equals zero at infinity. This requirement mathematically establishes:

$$\frac{1}{2}mv_e^2 - \frac{GMm}{R} = 0 \quad (26.2)$$

Solving this equation confirms the expression for v_e.

Trajectory Optimization for Gravity Assists

Gravity assist maneuvers, also referred to as gravity slingshots, are critical techniques for augmenting a spacecraft's velocity through the gravitational field of celestial bodies. They are implemented to conserve onboard fuel by exploiting the relative motion of planets.

1 Gravity Assist Mechanics

The gravity assist phenomenon uses the conservation of momentum in a frame of reference relative to the celestial body. Assuming a hyperbolic trajectory, velocity changes can be modeled using the vis-viva equation, adapted for two-body encounters:

$$v_f^2 = v_i^2 + 2GM\left(\frac{1}{r_p} - \frac{1}{r_i}\right) \quad (26.3)$$

where v_f and v_i are the final and initial velocities relative to the celestial body, r_p is the pericenter distance, and r_i the initial radial distance.

2 Velocity Increment through Swingby

The change in velocity Δv during a gravity assist is expressed by:

$$\Delta v = 2v_\infty \sin\left(\frac{\delta}{2}\right) \quad (26.4)$$

where v_∞ is the hyperbolic excess velocity and δ the angle of deflection, calculated from:

$$\delta = 2\arcsin\left(\frac{1}{1 + \frac{r_p v_\infty^2}{GM}}\right) \quad (26.5)$$

The above formulation provides insight into the achievable v, contingent upon approach parameters and the encountered body's gravitational characteristics.

Application in Mission Design

Deploying gravity assists optimizes interplanetary mission planning, enabling coverage of extensive distances or multiple planetary flybys. The optimization criteria focus on initial trajectory planning and precise navigation, incorporating perturbative effects due to mass distribution anomalies and solar radiation pressure.

1 Patched Conic Approximations

The patched conic approximation technique simplifies multi-body gravitational influences into sequential two-body problems. The initial heliocentric trajectory to the gravitational sphere of influence of the assisting body transitions to a hyperbolic trajectory, post-encounter return to heliocentric motion. The efficiency of this technique is enhanced by detailed numerical simulations utilizing optimization algorithms to minimize fuel consumption while meeting mission constraints.

Python Code Snippet

Below is a Python code snippet that encompasses the core computational elements related to escape velocity calculation, gravitational potential and kinetic energy balance, and trajectory optimization for gravity assists as discussed in this chapter.

```python
import numpy as np

def escape_velocity(G, M, R):
    '''
    Calculate the escape velocity from a celestial body.
    :param G: Gravitational constant.
    :param M: Mass of the celestial body.
    :param R: Radius of the celestial body.
    :return: Escape velocity.
    '''
    return np.sqrt(2 * G * M / R)

def gravity_assist_velocity_change(v_inf, rp, G, M):
    '''
    Calculate the velocity change from a gravity assist maneuver.
    :param v_inf: Hyperbolic excess velocity.
    :param rp: Pericenter distance of the trajectory.
    :param G: Gravitational constant.
    :param M: Mass of the planet being used for gravity assist.
```

```
    :return: Velocity change.
    '''
    delta = 2 * np.arcsin(1 / (1 + rp * v_inf**2 / (G * M)))
    return 2 * v_inf * np.sin(delta / 2)

def patched_conic_approximations(M, G, initial_velocity, r_initial,
↪   r_final):
    '''
    Simulate the patched conic approximation for interplanetary
    ↪   transfer.
    :param M: Mass of the celestial body being passed.
    :param G: Gravitational constant.
    :param initial_velocity: Initial velocity of the spacecraft.
    :param r_initial: Initial radial distance.
    :param r_final: Final radial distance.
    :return: Final velocity after gravity assist.
    '''
    v_i = initial_velocity
    delta_v = gravity_assist_velocity_change(v_i, r_initial, G, M)
    v_f = np.sqrt(v_i**2 + 2 * G * M * (1 / r_final - 1 /
    ↪   r_initial))
    return v_f + delta_v

# Constants
G = 6.67430e-11   # Gravitational constant (m^3 kg^-1 s^-2)
M_planet = 5.972e24   # Mass of Earth (kg)
R_planet = 6371000   # Radius of Earth (m)

# Calculations
v_escape = escape_velocity(G, M_planet, R_planet)
v_inf = 1e4   # Example hyperbolic excess velocity (m/s)
rp = 7000000   # Example pericenter distance (m)

# Example of velocity calculations using patched conic
↪   approximations
v_final = patched_conic_approximations(M_planet, G, v_inf, rp,
↪   R_planet)

# Output Results for Demonstration
print("Escape Velocity:", v_escape)
print("Velocity Change through Gravity Assist:",
↪   gravity_assist_velocity_change(v_inf, rp, G, M_planet))
print("Final Velocity after Patched Conic Approximation:", v_final)
```

This code defines several key functions necessary for understanding escape velocity calculations and the mechanics of gravity assists:

- `escape_velocity` computes the velocity required for an object to break free from a celestial body's gravitational influence.

- `gravity_assist_velocity_change` calculates the change in velocity achieved through a gravity assist maneuver, relying on key parameters such as hyperbolic excess velocity and pericenter distance.

- `patched_conic_approximations` simulates complex gravitational interactions in interplanetary missions using the patched conic approximation method.

The final block of code demonstrates these computations using example values and outputs the results for clarity and learning.

Chapter 27

Non-Linear Dynamics of Rocket Systems

Introduction to Non-Linear Dynamics

Non-linear dynamics encompass a broad class of behaviors in rocket systems resulting from equations of motion that are not linear with respect to their dependent variables or parameters. Such non-linearities can originate from aerodynamic forces, structural flexibility, propellant slosh, and control interactions. The mathematical representation of these systems requires advanced modeling techniques.

Equations of Motion

The foundation of analyzing non-linear dynamics in rocket systems begins with establishing the equations of motion. For a rigid body, the equations are typically derived from Euler's equations, which are expressed in their general form as:

$$\mathbf{I}\dot{\boldsymbol{\omega}} + \boldsymbol{\omega} \times (\mathbf{I}\boldsymbol{\omega}) = \mathbf{T} \tag{27.1}$$

where $\mathbf{I}$ denotes the inertia tensor, $\boldsymbol{\omega}$ represents the angular velocity vector, and $\mathbf{T}$ is the vector of external torques. These equations become non-linear due to the cross product term $\boldsymbol{\omega} \times (\mathbf{I}\boldsymbol{\omega})$.

Aerodynamic Non-Linearities

Non-linearities in aerodynamics are often described through empirical models or look-up tables. The aerodynamic forces and moments can be expressed as:

$$\mathbf{F}_a = \int_S \mathbf{p}\, dS$$
$$\mathbf{M}_a = \int_S (\mathbf{r} \times \mathbf{p})\, dS \quad (27.2)$$

where S is the surface area, $\mathbf{p}$ is the pressure distribution over the surface, and $\mathbf{r}$ is the position vector from the center of mass. The challenge arises in accurately modeling $\mathbf{p}$, which often depends on angle of attack and Mach number, both yielding non-linear relationships.

Structural Flexibility

Structural dynamics and flexibility introduce non-linear dynamics via coupling between rigid body and vibrational modes. The equation for coupled dynamics is given by:

$$\mathbf{M}\ddot{\mathbf{x}} + \mathbf{C}(\mathbf{x}, \dot{\mathbf{x}})\dot{\mathbf{x}} + \mathbf{K}(\mathbf{x})\mathbf{x} = \mathbf{F}(\mathbf{x}, \dot{\mathbf{x}}) \quad (27.3)$$

where $\mathbf{M}$ is the mass matrix, $\mathbf{C}$ is the damping matrix, $\mathbf{K}$ is the stiffness matrix, and $\mathbf{F}$ is the external force vector. Non-linearities manifest in the damping and stiffness matrices due to large displacements.

Propellant Slosh Dynamics

Propellant slosh introduces complex non-linear behaviors due to fluid-structure interactions. The slosh dynamics can be modeled using the pendulum analogy with added non-linear damping, expressed as:

$$\ddot{\theta} + c\dot{\theta} + \frac{g}{l}\sin\theta = 0 \quad (27.4)$$

where θ is the slosh angle, c is the damping coefficient, g is the acceleration due to gravity, and l is the length of the pendulum equivalent. The term $\sin\theta$ is inherently non-linear, leading to challenges in predictability and control.

Control System Interactions

Rocket control systems, including guidance, navigation, and control (GNC), often deal with complex non-linear behavior. These systems use state-space representations that involve non-linear functions:

$$\dot{\mathbf{x}} = \mathbf{f}(\mathbf{x}, \mathbf{u}, t)$$
$$\mathbf{y} = \mathbf{g}(\mathbf{x}, \mathbf{u}, t) \quad (27.5)$$

where $\mathbf{x}$ is the state vector, $\mathbf{u}$ is the input vector, $\mathbf{f}$ and $\mathbf{g}$ are non-linear functions, and $\mathbf{y}$ is the output vector. These require sophisticated control techniques such as feedback linearization or adaptive control to handle non-linearities effectively.

Numerical Methods for Non-Linear Systems

Non-linear dynamics in rocket systems often necessitate numeric integration techniques to solve complex ordinary differential equations (ODEs). Methods such as Runge-Kutta and predictor-corrector algorithms are standard. The implicit integration scheme for such systems is demonstrated as:

$$\mathbf{x}_{n+1} = \mathbf{x}_n + \Delta t \mathbf{f}(\mathbf{x}_n, \mathbf{u}_n, t_n) \quad (27.6)$$

where Δt is the time step. To improve convergence and stability in non-linear systems, adaptive step-size control and iterative solvers are employed.

Applications of Non-Linear Analysis

The application of non-linear dynamics analysis is paramount in the design and operation of rocket systems. From launch to landing, each phase encounters different non-linear phenomena, necessitating comprehensive modeling. The integration of such analyses into flight software ensures the reliability of trajectory corrections and stability margins across varying flight conditions.

Python Code Snippet

Below is a Python code snippet that implements the core mathematical components of non-linear dynamics in rocket systems, focusing on the calculations for rigid body motion, aerodynamic forces, structural flexibility, propellant slosh dynamics, control system interactions, and numerical integration.

```python
import numpy as np
from scipy.integrate import solve_ivp

# Define constants
g = 9.81  # acceleration due to gravity (m/s^2)
l = 1.0   # equivalent pendulum length (m)
mass_matrix = np.array([[1, 0], [0, 1]])  # Simplified mass matrix

# Euler's equations for rigid body dynamics
def eulers_equations(omega, I, torques):
    inertia_tensor = np.array(I)
    return np.linalg.inv(inertia_tensor) @ (torques -
        np.cross(omega, inertia_tensor @ omega))

# Aerodynamic forces calculation
def aerodynamic_forces(surface_pressure, surface_area,
    position_vector):
    force = np.trapz(surface_pressure * surface_area)
    moment = np.trapz(np.cross(position_vector, surface_pressure) *
        surface_area)
    return force, moment

# Structural flexibility dynamics
def structural_dynamics(t, x, mass_matrix, damping_matrix,
    stiffness_matrix, external_forces):
    # Assuming a linear damping and stiffness matrix for
    #   demonstration
    return np.linalg.inv(mass_matrix) @ (external_forces -
        damping_matrix @ x[1] - stiffness_matrix @ x[0])

# Propellant slosh modeled as a pendulum
def propellant_slosh(t, theta_c):
    theta, theta_dot = theta_c
    damping_coefficient = 0.05  # damping coefficient typical for
    #   slosh
    return [theta_dot, -damping_coefficient * theta_dot - (g/l) *
        np.sin(theta)]

# Control system non-linear state-space representation
def control_system_state(t, state, input_vector, f_func, g_func):
    # Define state derivative using non-linear functions f and g
    return f_func(state, input_vector, t)
```

```python
# Numerical integration for non-linear systems using Runge-Kutta
def integrate_nonlinear_system(dynamics_func, initial_conditions,
        time_span):
    sol = solve_ivp(dynamics_func, time_span, initial_conditions,
        method='RK45')
    return sol

# Example inputs for calculations
omega = [0.1, 0.2, 0.3]   # angular velocity vector
I = [1, 0.8, 0.9]         # inertia tensor values
torques = [0.5, 0.3, 0.2] # external torques

# Example aerodynamic calculation inputs
surface_pressure = np.array([101325, 100000, 99500])   # pressure
    distribution in Pascals
surface_area = np.array([1.2, 1.1, 1.0])               # surface area
    segment
position_vector = np.array([0.5, 0.4, 0.3])            # position
    vector from center of mass

# Initial conditions for structural dynamics
x0 = [0.0, 0.0]  # initial displacement and velocity

# Initial conditions for propellant slosh
theta_initial = [0.1, 0]  # initial angle and velocity

# Example usage: Rigid body dynamics
omega_dot = eulers_equations(omega, I, torques)
print('Angular acceleration:', omega_dot)

# Example usage: Aerodynamic forces
force, moment = aerodynamic_forces(surface_pressure, surface_area,
    position_vector)
print('Aerodynamic force:', force, 'Moment:', moment)

# Example usage: Structural flexibility
external_forces = [0.0, 0.0]
damping_matrix = np.array([[0.1, 0], [0, 0.1]])  # Simplified
    damping matrix
stiffness_matrix = np.array([[1, 0], [0, 1]])    # Simplified
    stiffness matrix
solution_structural = integrate_nonlinear_system(
    lambda t, x: structural_dynamics(t, x, mass_matrix,
        damping_matrix, stiffness_matrix, external_forces),
    x0, (0, 10)
)
print('Structural dynamics solution:', solution_structural.y)

# Example usage: Propellant slosh dynamics
solution_slosh = integrate_nonlinear_system(propellant_slosh,
    theta_initial, (0, 10))
print('Propellant slosh solution:', solution_slosh.y)
```

This code snippet defines the essential equations and models for understanding non-linear dynamics in rocket systems:

- `eulers_equations` function calculates angular acceleration based on Euler's equations for rigid body dynamics.

- `aerodynamic_forces` estimates the aerodynamic forces and moments using pressure distributions over the rocket surface.

- `structural_dynamics` models the dynamics of a flexible structure by integrating mass, damping, and stiffness matrices.

- `propellant_slosh` simulates the slosh dynamics using a pendulum analogy with non-linear damping.

- `integrate_nonlinear_system` performs numerical integration of ordinary differential equations using a Runge-Kutta method, applicable across different dynamic functions.

The code demonstrates these functions with examples, calculating dynamics for rigid body motion, aerodynamic forces, structural flexibility, and propellant slosh.

Chapter 28

Time of Flight Calculations

Introduction to Time of Flight

Time of flight calculations are pivotal in aerospace engineering for the optimization of trajectory planning and precise execution of docking sequences. The ability to predict the time a vehicle will spend in various phases of flight is crucial for mission success, ensuring synchronization with orbital dynamics and minimizing propellant consumption.

Fundamental Equations

The core equation governing time of flight is derived from integrating the velocity function over a specified trajectory. The basic form for time of flight T along a trajectory path from point A to point B is given by:

$$T = \int_{s_A}^{s_B} \frac{1}{v(s)} ds \qquad (28.1)$$

where $v(s)$ is the velocity as a function of path length s. For varying propulsion systems and mission profiles, this integration must consider changes in velocity due to thrust, gravitational forces, and drag.

Velocity and Acceleration Profiles

The velocity profile $v(t)$ of a vehicle is fundamental to determining the time of flight. Frequently, the profile can be segmented into different phases—such as boost, coast, and re-entry—each contributing uniquely to the overall flight time. For a constant acceleration scenario, the velocity is expressed as:

$$v(t) = v_0 + at \tag{28.2}$$

where v_0 is the initial velocity and a is the acceleration. The corresponding time of flight for constant acceleration is obtained through kinematic integration.

Trajectory Optimization Techniques

Optimizing trajectory for minimum time of flight involves solving variational problems, often employing calculus of variations or Pontryagin's maximum principle. The objective is to minimize the integral of time while satisfying dynamic and path constraints. The Hamiltonian for such a problem can be expressed as:

$$\mathcal{H}(t, \mathbf{x}, \mathbf{u}, \lambda) = 1 + \lambda^T f(\mathbf{x}, \mathbf{u}, t) \tag{28.3}$$

where $\mathbf{x}$ is the state vector, $\mathbf{u}$ is the control vector, and λ is the costate vector. Solving the optimal control problem yields a trajectory that minimizes time of flight while adhering to physical and environmental constraints.

Numerical Integration for Time of Flight

The complexity of integrating the time of flight equations necessitates numerical methods. The `Runge-Kutta` family of methods is frequently employed to solve the resulting initial value problems, particularly due to their balance of computational cost and accuracy. The general form of a `Runge-Kutta` method is:

$$x_{n+1} = x_n + h \sum_{i=1}^{s} b_i k_i \tag{28.4}$$

with stage evaluations:

$$k_i = f\left(t_n + c_i h, x_n + h\sum_{j=1}^{i-1} a_{ij} k_j\right) \qquad (28.5)$$

where h is the time step and a_{ij}, b_i, c_i are method-specific coefficients.

Docking Sequence Timing

Accurate timing of docking sequences in space missions requires the synchronization of the chaser and target vehicles. Closed-form solutions often become infeasible due to the nonlinear nature of orbital mechanics, requiring iterative algorithms to converge on a solution. The Lambert's problem framework provides a basis for calculating transfer times, assuming two-body motion between initial and final orbits.

$$\Delta T = \frac{\Delta\theta \cdot r_0 \cdot \sqrt{\frac{r_1}{\mu}}}{\sqrt{2}} \cdot \left(1 - \frac{e^2}{2}\right) \qquad (28.6)$$

where ΔT is the time of flight, $\Delta\theta$ the phase angle, r_0 and r_1 the radii of initial and final orbits, μ the standard gravitational parameter, and e the eccentricity of the transfer orbit.

Case Study: Rendezvous Mission

Examining the dynamics of a rendezvous mission highlights the practical applications of time of flight calculations. The case study entails a satellite performing a transfer maneuver to dock with a space station. By analyzing the thrust phases and coast arcs, the mission planners determine an optimal schedule balancing fuel efficiency and transfer time.

Vehicle dynamics during such maneuvers are modelled using the Clohessy-Wiltshire equations for relative motion in terms of the chaser's motion relative to the target. These equations inform the timing and thrust requirements necessary to achieve a successful dock within the predetermined window.

Python Code Snippet

Below is a Python code snippet that encompasses the core computational elements for time of flight calculations, including velocity and trajectory optimization, numerical integration, and docking sequence timing.

```python
import numpy as np
from scipy.integrate import solve_ivp

def time_of_flight(velocity_function, s_A, s_B):
    '''
    Calculate the time of flight between two points.
    :param velocity_function: Function describing velocity as a
       function of path.
    :param s_A: Start point.
    :param s_B: End point.
    :return: Time of flight.
    '''
    func = lambda s: 1.0 / velocity_function(s)
    time, _ = np.trapz(func(np.linspace(s_A, s_B, num=1000)))
    return time

def constant_acceleration_velocity(v0, a, t):
    '''
    Calculate velocity with constant acceleration.
    :param v0: Initial velocity.
    :param a: Acceleration.
    :param t: Time.
    :return: Velocity.
    '''
    return v0 + a * t

def runge_kutta_integration(f, y0, t_span, h):
    '''
    Runge-Kutta numerical integration.
    :param f: Function to integrate.
    :param y0: Initial conditions.
    :param t_span: Time span.
    :param h: Time step.
    :return: Integration result.
    '''
    return solve_ivp(f, t_span, y0, method='RK45', max_step=h)

def trajectory_optimization(state_vector, control_vector):
    '''
    Optimize trajectory using calculus of variations.
    :param state_vector: Initial state.
    :param control_vector: Control inputs.
    :return: Optimized trajectory.
    '''
```

```
# Placeholder for actual optimization algorithm
optimized_path = state_vector + np.dot(control_vector,
↪    np.random.random(len(state_vector)))
return optimized_path

def lamberts_problem(T, delta_theta, r0, r1, mu, e):
    '''
    Calculate time of flight using Lambert's problem.
    :param T: Time of flight.
    :param delta_theta: Phase angle.
    :param r0: Radius of initial orbit.
    :param r1: Radius of final orbit.
    :param mu: Gravitational parameter.
    :param e: Eccentricity of orbit.
    :return: Time of flight.
    '''
    return (delta_theta * r0 * np.sqrt(r1 / mu) / np.sqrt(2)) * (1 -
    ↪    e**2 / 2)

# Example usage
velocity_fn = lambda s: constant_acceleration_velocity(100, 9.81, s)
tof = time_of_flight(velocity_fn, 0, 1000)

initial_state = np.array([0, 0, 0])
control_inputs = np.array([1.5, -0.5, 0.5])
optimized = trajectory_optimization(initial_state, control_inputs)

print("Time of Flight:", tof)
print("Optimized Trajectory:", optimized)
```

This code defines several key functions necessary for calculating the time of flight in rocket systems:

- `time_of_flight` function computes the time it takes for a vehicle to travel between two points given a velocity function.

- `constant_acceleration_velocity` calculates velocity under constant acceleration conditions.

- `runge_kutta_integration` performs numerical integration using the Runge-Kutta method for solving differential equations.

- `trajectory_optimization` provides a placeholder for optimizing trajectories using control inputs and state vectors.

- `lamberts_problem` exemplifies the use of Lambert's problem for calculating time of flight in orbital mechanics.

The final block of code provides example computations using these functions.

Chapter 29

Fuel Optimal Control Techniques

Introduction to Fuel Optimization

The field of fuel optimal control presents a vital aspect in the design and execution of aerospace missions, prioritizing efficiency in propellant usage while achieving mission objectives. These optimization techniques are critical in shaping the trajectory and control strategies of aerospace vehicles, directly impacting the payload capacity and mission duration.

Optimal Control Formulations

In fuel optimal control, a primary goal is minimizing the integral of a cost function representing fuel consumption, often expressed as:

$$J = \int_{t_0}^{t_f} L(\mathbf{x}(t), \mathbf{u}(t), t)\, dt \qquad (29.1)$$

where J is the performance index, $\mathbf{x}(t)$ the state vector, $\mathbf{u}(t)$ the control vector, and L the Lagrangian associated with fuel consumption.

Dynamic Equations and Constraints

The motion of the vehicle is governed by a set of differential equations:

$$\dot{\mathbf{x}} = f(\mathbf{x}, \mathbf{u}, t) \tag{29.2}$$

subject to boundary conditions and path constraints expressed as:

$$\mathbf{g}(\mathbf{x}(t), \mathbf{u}(t), t) \leq 0 \tag{29.3}$$

Where **g** represents the inequality constraints capturing physical limitations and safety requirements.

Applications of Pontryagin's Maximum Principle

Pontryagin's Maximum Principle (PMP) provides a necessary condition for optimality by declaring that the Hamiltonian $\mathcal{H}$ must be maximized with respect to the control variable **u**:

$$\mathcal{H} = L(\mathbf{x}, \mathbf{u}, t) + \lambda^T f(\mathbf{x}, \mathbf{u}, t) \tag{29.4}$$

where λ is the costate vector satisfying the adjoint equation:

$$\dot{\lambda} = -\frac{\partial \mathcal{H}}{\partial \mathbf{x}} \tag{29.5}$$

Fuel Optimal Thrust Arcs

In scenarios involving thrust arcs, the determination of optimal thrust profiles is crucial. Discrete switching times between thrust and coast phases must satisfy:

$$\delta \mathcal{H} = 0, \quad \text{with constraints on } \mathbf{u} \tag{29.6}$$

The optimal thrust profile enables minimization of the performance index J while adhering to dynamic equation constraints.

Numerical Methods for Solving Optimal Control Problems

Numerical methods such as the `Collocation` method and `Direct Multiple Shooting` are often employed. These techniques transform the control problem into a nonlinear programming problem (NLP), solvable using optimization algorithms.

Direct multiple shooting subdivides the time horizon and solves for both states and controls iteratively, maintaining continuity and dynamical accuracy with:

$$\text{minimize} \sum_k J_k = \sum_k \int_{t_k}^{t_{k+1}} L(\mathbf{x}(t), \mathbf{u}(t), t) \, dt \qquad (29.7)$$

subject to the satisfaction of the vehicle dynamics and constraints.

Case Study: Low Thrust Trajectory Optimization

For missions requiring low-thrust propulsion systems, optimal control strategies are indispensable. In such cases, the efficiency of electric propulsion is maximized by continuously varying thrust angle and magnitude, evaluated through an optimal control framework.

The equations of motion might be updated to reflect the low thrust consideration:

$$\dot{v} = \frac{T}{m} - g + \frac{\mathbf{F}_d}{m} \qquad (29.8)$$

where T is the thrust, m the mass of the vehicle, g the gravitational acceleration, and $\mathbf{F}_d$ the drag force. The selection of optimal thrust vectors is achieved by leveraging the control algorithms articulated in preceding sections.

Python Code Snippet

Below is a Python code snippet that implements key computational concepts from this chapter on fuel optimal control techniques including numerical resolution using direct multiple shooting, Pontryagin's Maximum Principle applications, and low thrust trajectory optimization.

```
import numpy as np
from scipy.optimize import minimize

def lagrangian(x, u, t):
    '''
    Define the Lagrangian function for fuel consumption.
    :param x: State vector.
    :param u: Control vector.
    :param t: Time.
    :return: The Lagrangian value.
    '''
    return np.linalg.norm(u)

def dynamics(x, u, t):
    '''
    Vehicle dynamics equations
    :param x: State vector.
    :param u: Control vector.
    :param t: Time.
    :return: Derivative of the state vector.
    '''
    return np.array([
        x[1],
        u[0] / x[2], # Simple thrust model
        -9.81,       # Constant gravity term
    ])

def pontryagin_maximum(t, x, lambda_, control_bounds):
    '''
    Apply Pontryagin's Maximum Principle to find optimal control.
    :param t: Current time.
    :param x: Current state.
    :param lambda_: Costate vector.
    :param control_bounds: Control limits.
    :return: Optimal control vector.
    '''
    def hamiltonian(u):
        return lagrangian(x, u, t) + np.dot(lambda_, dynamics(x, u,
        ↪ t))

    opt_result = minimize(lambda u: -hamiltonian(u),
    ↪ np.zeros_like(x), bounds=control_bounds)
    return opt_result.x

def direct_multiple_shooting(t0, tf, x0, n_segments, control_bounds,
↪ n_iterations):
    '''
```

```
    Solve the optimal control problem using Direct Multiple
 ↪   Shooting.
    :param t0: Initial time.
    :param tf: Final time.
    :param x0: Initial state vector.
    :param n_segments: Number of time segments.
    :param control_bounds: Bounds for control vector.
    :param n_iterations: Number of optimization iterations.
    :return: Optimal state and control trajectories.
    '''
    time = np.linspace(t0, tf, n_segments + 1)
    x_trajectory = np.zeros((n_segments + 1, len(x0)))
    u_trajectory = np.zeros((n_segments, len(x0)))
    x_trajectory[0] = x0

    for iteration in range(n_iterations):
        for k in range(n_segments):
            u_trajectory[k] = pontryagin_maximum(time[k],
 ↪          x_trajectory[k], np.zeros_like(x0), control_bounds)
            h = time[k + 1] - time[k]
            x_trajectory[k + 1] = x_trajectory[k] + h *
 ↪          dynamics(x_trajectory[k], u_trajectory[k], time[k])

    return x_trajectory, u_trajectory

def low_thrust_trajectory_optimization(x0, tf):
    '''
    Optimize a low thrust trajectory.
    :param x0: Initial state vector.
    :param tf: Final time.
    :return: Optimized trajectory for low-thrust propulsion.
    '''
    n_segments = 100
    control_bounds = [(-10, 10), (-10, 10), (-10, 10)]
    n_iterations = 10

    return direct_multiple_shooting(0, tf, x0, n_segments,
 ↪  control_bounds, n_iterations)

# Initial conditions
x0 = np.array([0, 0, 500])   # Initial position, speed, and mass
tf = 300   # Final time

# Calculate optimized trajectory
x_traj, u_traj = low_thrust_trajectory_optimization(x0, tf)

print("Optimized States:", x_traj)
print("Optimized Controls:", u_traj)
```

This code snippet covers essential functions and algorithms for resolving fuel optimal control challenges within aerospace engineering:

- `lagrangian` function defines the cost function associated with fuel consumption.

- `dynamics` describes the motion of the vehicle under thrust and gravitational forces.

- `pontryagin_maximum` applies Pontryagin's Maximum Principle to obtain optimal controls.

- `direct_multiple_shooting` provides a structured approach to tackle optimal control problems numerically.

- `low_thrust_trajectory_optimization` simulates the conditions of a low-thrust propulsion system for trajectory adjustment.

The example showcases how to compute optimized states and controls for a hypothetical aerospace mission by employing direct multiple shooting and Pontryagin's Maximum Principle.

Chapter 30

Kalman Filter Applications in Navigation

Introduction to Kalman Filters in Aerospace Navigation

The Kalman Filter (KF) stands as a cornerstone in modern navigation systems for aerospace applications. Its ability to estimate the state of a dynamic system from noisy measurements makes it invaluable for rockets and spacecraft navigating variable space conditions. The Kalman Filter combines system modeling with statistically optimal estimation to improve trajectory accuracy and reliability. In aerospace engineering, the versatility of the Kalman Filter allows for integration with various sensor inputs to correct and predict system states, offering enhanced precision and robustness.

Mathematical Foundations of the Kalman Filter

The Kalman Filter operates in a recursive manner, employing a two-phase process of prediction and update. The discrete-time KF model can be represented mathematically by the following state-space equations:

$$\mathbf{x}_{k+1} = \mathbf{F}_k \mathbf{x}_k + \mathbf{B}_k \mathbf{u}_k + \mathbf{w}_k \tag{30.1}$$

$$\mathbf{z}_k = \mathbf{H}_k \mathbf{x}_k + \mathbf{v}_k \tag{30.2}$$

where $\mathbf{x}_k$ denotes the state vector at time k, $\mathbf{u}_k$ is the control input, $\mathbf{w}_k$ is the process noise, typically assumed to have a normal distribution with covariance $\mathbf{Q}_k$. The measurement model is given by the second equation, where $\mathbf{z}_k$ is the measurement vector, and $\mathbf{v}_k$ represents the measurement noise, assumed to be normally distributed with covariance $\mathbf{R}_k$.

Prediction and Update Equations

The filter prediction step is designed to forecast the next state and its uncertainty:

$$\hat{\mathbf{x}}_{k|k-1} = \mathbf{F}_{k-1}\hat{\mathbf{x}}_{k-1|k-1} + \mathbf{B}_{k-1}\mathbf{u}_{k-1} \tag{30.3}$$

$$\mathbf{P}_{k|k-1} = \mathbf{F}_{k-1}\mathbf{P}_{k-1|k-1}\mathbf{F}_{k-1}^T + \mathbf{Q}_{k-1} \tag{30.4}$$

where $\hat{\mathbf{x}}_{k|k-1}$ is the predicted state estimate, and $\mathbf{P}_{k|k-1}$ is the predicted estimate covariance.

During the update phase, innovative data from measurements refine the prediction:

$$\mathbf{K}_k = \mathbf{P}_{k|k-1}\mathbf{H}_k^T \left(\mathbf{H}_k \mathbf{P}_{k|k-1}\mathbf{H}_k^T + \mathbf{R}_k\right)^{-1} \tag{30.5}$$

$$\hat{\mathbf{x}}_{k|k} = \hat{\mathbf{x}}_{k|k-1} + \mathbf{K}_k \left(\mathbf{z}_k - \mathbf{H}_k \hat{\mathbf{x}}_{k|k-1}\right) \tag{30.6}$$

$$\mathbf{P}_{k|k} = (\mathbf{I} - \mathbf{K}_k \mathbf{H}_k) \mathbf{P}_{k|k-1} \tag{30.7}$$

where $\mathbf{K}_k$ is the Kalman gain, which determines the weighting of the prediction versus measurement. $\hat{\mathbf{x}}_{k|k}$ and $\mathbf{P}_{k|k}$ are the updated state estimate and updated estimate covariance, respectively.

Applications in Spacecraft Navigation

The Kalman Filter's adaptability to different navigation systems makes it suitable for precision maneuvers in space. Onboard sensors, such as accelerometers and gyroscopes, alongside external aids from ground stations or satellites, feed the KF with necessary measurements. The filter's recursive algorithms update the spacecraft's trajectory estimates continually, even amidst disruption from variable gravitational fields and other dynamic space environments.

Extended and Unscented Kalman Filters

For non-linear systems, extensions of the Kalman Filter are often utilized, such as the Extended Kalman Filter (EKF) and Unscented Kalman Filter (UKF). The EKF linearizes the non-linear model at each timestep, using Jacobian matrices:

$$\mathbf{F}_k = \left.\frac{\partial f}{\partial \mathbf{x}}\right|_{\hat{\mathbf{x}}_{k|k-1}, \mathbf{u}_k} \tag{30.8}$$

$$\mathbf{H}_k = \left.\frac{\partial h}{\partial \mathbf{x}}\right|_{\hat{\mathbf{x}}_{k|k-1}} \tag{30.9}$$

Alternatively, the UKF utilizes a deterministic sampling method to propagate means and covariances through non-linear transformations, avoiding linearization errors associated with the EKF. The selection of method depends on system complexity and the degree of non-linearity.

Implementation in Rocket Guidance Systems

The implementation of Kalman Filter algorithms in rocket guidance systems aids in the autonomous control of launch vehicle trajectories and orbital insertions. Integration of such systems necessitates careful design considerations to accommodate the high dynamic range and potential measurement biases due to severe environmental conditions. The continual refinement of state vectors ensures that corrections to the navigational pathway are executed with minimal latency.

Python Code Snippet

Below is a Python code snippet that encompasses the core computational elements for the implementation of the Kalman Filter in aerospace navigation systems, including state prediction, update calculations, and handling of sensor data.

```python
import numpy as np

def kalman_filter(F, B, H, Q, R, x0, P0, measurements,
                  control_inputs):
    '''
    Implement the Kalman Filter for linear systems.
    :param F: State transition matrix.
    :param B: Control input matrix.
    :param H: Observation matrix.
    :param Q: Process noise covariance matrix.
    :param R: Measurement noise covariance matrix.
    :param x0: Initial state estimate.
    :param P0: Initial estimate covariance.
    :param measurements: Measurement data.
    :param control_inputs: Control input data.
    :return: Filtered state estimates.
    '''
    x = x0
    P = P0

    estimates = [x]

    for k in range(len(measurements)):
        # Prediction step
        x_pred = F @ x + B @ control_inputs[k]
        P_pred = F @ P @ F.T + Q

        # Update step
        y = measurements[k] - H @ x_pred
        S = H @ P_pred @ H.T + R
        K = P_pred @ H.T @ np.linalg.inv(S)   # Kalman gain

        x = x_pred + K @ y
        P = (np.eye(len(P)) - K @ H) @ P_pred

        estimates.append(x)

    return estimates

# System parameters
F = np.array([[1, 1], [0, 1]])    # State transition matrix
B = np.array([[0.5], [1]])        # Control input matrix
H = np.array([[1, 0]])            # Observation matrix
Q = np.array([[1, 0], [0, 1]])    # Process noise covariance
```

```
R = np.array([[1]])              # Measurement noise covariance
x0 = np.array([0, 0])            # Initial state estimate
P0 = np.eye(2)                   # Initial estimate covariance

# Dummy data
measurements = [1, 2, 3, 4, 5]
control_inputs = [0, 0, 0, 0, 0]

# Execute Kalman Filter
state_estimates = kalman_filter(F, B, H, Q, R, x0, P0, measurements,
    control_inputs)

for i, est in enumerate(state_estimates):
    print(f"Estimate at step {i}: {est}")
```

This code defines the key functions and parameters required for implementing the Kalman Filter in aerospace navigation:

- `kalman_filter` function operates the recursive process of predicting and updating state estimates using system dynamics and measurements.

- Inputs to this function include matrices such as F (state transition) and H (observation), as well as noise covariances Q and R.

- The filter is initialized with an initial state estimate, x0, and initial estimate covariance, P0.

- The `measurements` and `control_inputs` lists provide time series data fed into the filter at each step.

- The final block of code executes the `kalman_filter` with dummy data and prints out the sequence of state estimates.

The Python snippet provides a practical demonstration of applying the Kalman Filter, which is central to improving navigation accuracy in aerospace systems.

Chapter 31

Monte Carlo Simulations in Propulsion

Introduction to Monte Carlo Simulations

Monte Carlo simulations play a critical role in aerospace engineering for the analysis of uncertainty and risk within propulsion systems. By leveraging random sampling and statistical methods, these simulations help quantify variabilities in complex systems that are subject to a multitude of uncertain parameters. The methodology offers insights into performance distributions rather than deterministic outcomes, making it particularly useful for propulsion calculations.

In aerospace propulsion, Monte Carlo simulations assist in modeling the stochastic nature of inputs and their propagation through system models to predict outputs under varied conditions. Variability sources such as material properties, geometric tolerances, and environmental conditions are incorporated into the simulations to evaluate their collective effects on propulsion system reliability.

Fundamental Principles of Monte Carlo Methods

Monte Carlo methods simulate random variables to model probabilistic systems and phenomena. The process typically begins with the definition of input probability distributions for uncertain parameters. These distributions are sampled iteratively to provide a range of possible inputs to deterministic models, generating a distribution of outputs.

Consider a propulsion parameter X characterized by a probability density function (PDF) $f_X(x)$. The Monte Carlo methodology samples from $f_X(x)$, providing inputs to a model $g(x)$ that transforms them into outputs $Y = g(X)$. The algorithm can be outlined as follows:

1. Define the input uncertainty model: $X \sim \text{PDF}(x)$. 2. Sample x_i from $f_X(x)$. 3. Compute $y_i = g(x_i)$. 4. Repeat steps 2-3 for $i = 1, 2, \ldots, N$. 5. Analyze $\{y_1, y_2, \ldots, y_N\}$ to estimate statistical properties.

Application to Propulsion System Uncertainties

Propulsion systems are influenced by numerous uncertain parameters impacting thrust, efficiency, and structural integrity. These include variations in fuel properties, combustion chamber pressures, and nozzle geometry deviations. Monte Carlo simulations are well-suited to assess how such uncertainties affect system outputs.

The transformed state equation for a given propulsion model can be expressed as:

$$Y = g(X_1, X_2, \ldots, X_n, \gamma)$$

where $X_1, X_2, \ldots, X_n$ are input random variables, and γ represents deterministic inputs. Through simulations, distributions of Y, such as the thrust output, can be obtained. This facilitates probabilistic evaluations of performance against defined specifications.

Risk Assessment via Monte Carlo Simulations

Risk assessment in propulsion involves quantifying the probability of failure or underperformance due to uncertainties. Monte Carlo methods offer a framework to compute the probability distribution of failure modes and critical performance thresholds.

Consider a system with a failure criterion:

$$R(X) \leq R_{\texttt{crit}}$$

where $R(X)$ is a response function dependent on input random variables and $R_{\texttt{crit}}$ is the critical performance threshold. Using Monte Carlo simulations, the failure probability P_f is estimated as:

$$P_f = \frac{\texttt{Number of Failures}}{N}$$

where failures are counted when $R(X_i) \leq R_{\texttt{crit}}$ for each simulated run. The fidelity of such risk assessments is primarily a function of the sample size N.

Mathematical Implementation of Monte Carlo Methods

Monte Carlo simulations fundamentally rely on mathematical constructs such as random number generation, statistical convergence, and variance reduction techniques. Pseudorandom number generators (RNG) initialize the sampling process, typically leveraging algorithms like Mersenne Twister for high-quality randomness.

Variance reduction techniques such as `importance sampling`, `stratified sampling`, and `antithetic variates` enhance simulation efficiency by minimizing the error in output estimations without increasing sample size N.

System Reliability Studies

Reliability studies in propulsion systems leverage Monte Carlo simulations to calculate the likelihood of maintaining operational conditions over time. By simulating various component life models and applying fatigue and wear laws, these studies can deduce mean time to failure and other reliability metrics under operational uncertainties.

The reliability function $R(t)$ characterizes system survivability over time t, expressed as:

$$R(t) = \exp\left(-\int_0^t \lambda(\tau)d\tau\right)$$

where $\lambda(\tau)$ denotes the failure rate, often a random variable within Monte Carlo frameworks representing degradation processes.

Python Code Snippet

Below is a Python code snippet that encompasses the core computational elements of Monte Carlo simulations in propulsion including the sampling of input distributions, calculation of outputs, risk assessment, and reliability evaluation.

```
import numpy as np

def monte_carlo_simulation(f_X, g, N, R_crit):
    '''
    Perform Monte Carlo simulation for propulsion system.
    :param f_X: Probability density function for input distribution.
    :param g: Deterministic model function transforming inputs to
        outputs.
    :param N: Number of simulations.
    :param R_crit: Critical performance threshold for failure.
    :return: Output distribution, failure probability.
    '''
    outputs = []
    failures = 0

    # Monte Carlo sampling and transformation
    for _ in range(N):
        # Sample input from the probability distribution
        x_i = f_X()
        # Calculate output using the deterministic function
        y_i = g(x_i)
        outputs.append(y_i)
        # Assess failure condition
        if y_i <= R_crit:
            failures += 1

    # Calculate failure probability
    P_f = failures / N
    return np.array(outputs), P_f
```

```python
def random_engine_input():
    '''
    Example random input distribution for an engine parameter.
    :return: Sampled engine input parameter.
    '''
    # Example: Assume the parameter follows a normal distribution
    return np.random.normal(loc=100, scale=5)

def propulsion_model(x):
    '''
    Deterministic propulsion model.
    :param x: Engine input parameter.
    :return: Calculated thrust output.
    '''
    # Example transformation function
    return x * 2.5 - 10

def failure_threshold():
    '''
    Define system failure threshold.
    :return: Critical threshold value.
    '''
    return 200

# Parameters for Monte Carlo Simulation
N = 10000  # Number of simulations
outputs, failure_probability = monte_carlo_simulation(
    f_X=random_engine_input,
    g=propulsion_model,
    N=N,
    R_crit=failure_threshold()
)

print("Simulation Outputs:", outputs)
print("Failure Probability:", failure_probability)

def system_reliability(N, lambda_rate):
    '''
    Evaluate system reliability using a stochastic failure rate.
    :param N: Lifetime simulations.
    :param lambda_rate: Failure rate distribution function.
    :return: Reliability over time.
    '''
    random_failures = lambda_rate(size=N)
    survival_probabilities = np.exp(-np.cumsum(random_failures))
    return survival_probabilities

def failure_rate():
    '''
    Example failure rate distribution.
    :return: Sampled failure rate.
    '''
```

```
# Example: Assume a constant failure rate with added noise
return np.random.uniform(low=0.01, high=0.02)

# Evaluate reliability
reliability = system_reliability(N=1000, lambda_rate=failure_rate)
print("System Reliability:", reliability)
```

This code defines several key functions necessary for executing Monte Carlo simulations specific to propulsion systems:

- `monte_carlo_simulation` performs the simulation by sampling the input distribution and calculating outputs, also determining failure probability against a set threshold.

- `random_engine_input` provides a mechanism for sampling random input values representing possible engine conditions.

- `propulsion_model` represents the deterministic transformation function used to compute thrust from inputs.

- `failure_threshold` states the critical threshold for determining failure.

- `system_reliability` calculates the reliability of the system over time using a specified failure rate distribution.

- `failure_rate` gives an example input for the failure rate distribution, accounting for stochastic variations.

Chapter 32

Benefits of Multi-Stage Rockets

Introduction to Multi-Stage Rockets

The concept of multi-stage rockets provides a robust framework for optimizing payload capacity and enhancing velocity increments. By discarding parts of the vehicle during flight, each subsequent stage takes advantage of lower mass, thereby making more efficient use of propellant. This section provides a foundational understanding of why multi-staging is employed in rocket propulsion systems. The primary benefit is the reduction of exponential growth in propellant mass as highlighted by the rocket equation.

Tsiolkovsky Rocket Equation for Multi-Stage Analysis

The Tsiolkovsky Rocket Equation is revisited in the context of multi-staging. For a single-stage rocket, the velocity increment Δv is calculated using:

$$\Delta v = v_e \ln\left(\frac{m_0}{m_f}\right)$$

where v_e is the effective exhaust velocity, m_0 is the initial mass, and m_f is the final mass. For a n-stage vehicle, the total increment in velocity ΔV is expressed as the sum of contributions from all stages:

$$\Delta V = \sum_{i=1}^{n} v_{e,i} \ln\left(\frac{m_{0,i}}{m_{f,i}}\right)$$

This equation demonstrates how each stage i contributes to the overall velocity increment.

Payload Fraction Optimizations

The payload fraction, defined as the ratio of payload mass m_L to the initial mass of the rocket m_0, is a key metric in assessing rocket design. This can be expressed mathematically for an n-stage rocket as:

$$\text{Payload Fraction} = \frac{m_L}{m_0} = \prod_{i=1}^{n}\left(\frac{m_{f,i}}{m_{0,i}}\right)$$

Here, $m_{0,i}$ and $m_{f,i}$ are the initial and final masses of each stage including structural components. Optimizing the payload fraction involves iterating over stage-specific parameters to maximize this ratio while achieving the desired velocity increments.

Staging Strategies and Configuration Analysis

Optimal multi-stage rocket designs are evaluated through various staging configurations like parallel and serial staging. Serial staging involves one stage firing after the exhaustion of the previous, while parallel includes the simultaneous operation of multiple stages.

For efficiency analysis, consider the staging mass m_s that gets discarded. The mass ratio R_s of a specific stage, assuming sequential staging, is defined by:

$$R_s = \frac{m_0}{m_f + m_s}$$

The performance is often depicted through a trade-off between the mass ratio and specific impulse I_{sp}. Parallel staging configurations may implement an optimization strategy referred to as 'parallel burn efficiency' defined by:

$$\eta_{\text{parallel}} = \frac{\Delta V_{\text{parallel}}}{\Delta V_{\text{ideal}}}$$

Where $\Delta V_{\text{parallel}}$ and ΔV_{ideal} represent the real and ideal velocities achievable via parallel staging configurations.

Velocity Increment Considerations and Stage Coupling

The distribution of total velocity increment ΔV among multiple stages involves solving optimization problems, particularly in Stage-coupled systems, where achieved ΔV_i for each stage should collectively equal the mission requirement $\Delta V_{\text{mission}}$:

$$\Delta V_{\text{mission}} = \sum_{i=1}^{n} \Delta V_i$$

For practicality, this involves complex algorithms that assess individual stage optimization within defined constraints such as total mass and thrust limitations. Performance by thrust at different mission phases is critical to this computation, often employing constraint optimization techniques for achieving optimal stage coupling.

Advanced Multi-Stage System Design

Incorporating advanced optimization techniques, such as genetic algorithms and constraint optimization, into system designs are crucial for establishing highly efficient multi-stage rockets. These algorithms optimize structural dynamics, propellant distribution, and other crucial parameters to achieve mission goals while minimizing costs.

Overall, mathematical frameworks employed in multi-stage rocket configurations inherently rely on detailed simulations and computational models that encapsulate realistic flight dynamics. Final stage performance metrics are heavily influenced by designed simulations that extrapolate real-world conditions into the computational workflow, ensuring reliability and efficiency throughout the rocket's trajectory.

Python Code Snippet

Below is a Python code snippet that encompasses the core computational elements for multi-stage rocket analysis, including calculations based on the Tsiolkovsky Rocket Equation, payload fraction optimization, and staging strategies.

```python
import numpy as np

def tsiolkovsky_velocity_increment(ve, m0, mf):
    '''
    Calculate the velocity increment using Tsiolkovsky Rocket
    ↪ Equation for a single stage.
    :param ve: Effective exhaust velocity.
    :param m0: Initial mass.
    :param mf: Final mass.
    :return: Velocity increment (delta v).
    '''
    return ve * np.log(m0 / mf)

def multi_stage_velocity_increment(ve_stages, m0_stages, mf_stages):
    '''
    Calculate the total velocity increment for a multi-stage rocket.
    :param ve_stages: List of effective exhaust velocities for each
    ↪ stage.
    :param m0_stages: List of initial masses for each stage.
    :param mf_stages: List of final masses for each stage.
    :return: Total velocity increment (delta V).
    '''
    return sum(ve * np.log(m0 / mf) for ve, m0, mf in zip(ve_stages,
    ↪ m0_stages, mf_stages))

def payload_fraction(ml, m0_stages, mf_stages):
    '''
    Calculate the payload fraction for a multi-stage rocket.
    :param ml: Payload mass.
    :param m0_stages: List of initial masses for each stage.
    :param mf_stages: List of final masses for each stage.
    :return: Payload fraction.
    '''
    total_mass_ratio = np.prod([mf / m0 for m0, mf in zip(m0_stages,
    ↪ mf_stages)])
    return ml / (m0_stages[0] * total_mass_ratio)

def stage_mass_ratio(m0, mf, ms):
    '''
    Calculate the mass ratio for a specific stage.
    :param m0: Initial mass of the stage.
    :param mf: Final mass of the stage.
    :param ms: Staging mass of the stage.
    :return: Mass ratio.
```

```python
    '''
    return m0 / (mf + ms)

def parallel_burn_efficiency(delta_v_parallel, delta_v_ideal):
    '''
    Calculate the parallel burn efficiency for rocket staging.
    :param delta_v_parallel: Real velocity achieved through parallel
     ↪  staging.
    :param delta_v_ideal: Ideal velocity increment without staging
     ↪  losses.
    :return: Parallel burn efficiency.
    '''
    return delta_v_parallel / delta_v_ideal

# Example values for a three-stage rocket
ve_stages = [3000, 3500, 4000]  # Effective exhaust velocities in
 ↪  m/s
m0_stages = [50000, 20000, 8000]   # Initial masses in kg
mf_stages = [30000, 10000, 4000]   # Final masses in kg
payload_mass = 1000   # Payload mass in kg

# Calculating the total velocity increment for a multi-stage rocket
total_delta_v = multi_stage_velocity_increment(ve_stages, m0_stages,
 ↪  mf_stages)
print("Total Delta V:", total_delta_v)

# Calculating the payload fraction
payload_frac = payload_fraction(payload_mass, m0_stages, mf_stages)
print("Payload Fraction:", payload_frac)

# Example staging mass and calculation of mass ratio
staging_mass = 1000   # Example staging mass in kg
mass_ratio_stage_1 = stage_mass_ratio(m0_stages[0], mf_stages[0],
 ↪  staging_mass)
print("Mass Ratio for Stage 1:", mass_ratio_stage_1)

# Example parallel burn efficiency calculation
delta_v_parallel = 7900   # Example real velocity in m/s
delta_v_ideal = 8000   # Example ideal velocity in m/s
parallel_eff = parallel_burn_efficiency(delta_v_parallel,
 ↪  delta_v_ideal)
print("Parallel Burn Efficiency:", parallel_eff)
```

This code defines several key functions necessary for performing core analysis of multi-stage rockets:

- `tsiokolvsky_velocity_increment` function calculates the velocity increment for a single stage using the Tsiolkovsky Rocket Equation.
- `multi_stage_velocity_increment` computes the total velocity increment for a multi-stage rocket.

- `payload_fraction` determines the payload fraction, assessing the mass efficiency of different rocket stages.

- `stage_mass_ratio` calculates the mass ratio of a specific stage based on its initial, final, and staging mass.

- `parallel_burn_efficiency` evaluates the efficiency of parallel staging through real versus ideal velocity increment.

To demonstrate the calculations, example values are provided for a three-stage rocket, revealing the potential performance metrics of multi-stage rocket configurations.

Chapter 33

Geosynchronous and Polar Orbits

Introduction to Orbital Dynamics

Within the field of astrodynamics, geosynchronous and polar orbits offer distinct strategic benefits for satellite deployment, addressing both longitudinal station-keeping and global coverage requirements. The differential orbital mechanics underlying these two types of orbits necessitate precise calculations and careful maneuver planning. This chapter will address the foundational methodologies involved in achieving maximum efficacy for satellites operating in these orbits.

Geosynchronous Orbit Calculations

A geosynchronous orbit (GEO) is characterized by an orbital period that matches the Earth's rotational period. Key parameters such as the semi-major axis, a, are critical to achieving synchrony. The relationship is defined by:

$$T = 2\pi \sqrt{\frac{a^3}{\mu}}$$

where T is the orbital period and μ is the Earth's standard gravitational parameter, approximately $3.986 \times 10^{14}\,\text{m}^3/\text{s}^2$.

For a geosynchronous satellite, $T = 86400\,\text{s}$. Solving for a yields:

$$a = \left(\frac{\mu T^2}{4\pi^2}\right)^{1/3}$$

The spherical geometry of Earth introduces perturbative forces over time, particularly the oblateness of Earth quantified by the J_2 harmonic. Modifications to the orbital inclination, i, and right ascension of the ascending node, Ω, are often accounted for using Lagrange's planetary equations or numerical integration with perturbation corrections.

Station-Keeping in Geosynchronous Orbits

To maintain geostationary properties, station-keeping maneuvers are periodically executed. The radial distance Δr, dictated by gravitation and atmospheric drag, requires compensatory thrust applications calculated by:

$$\Delta V = \sqrt{\mu}\left|\sqrt{\frac{2}{r}} - \sqrt{\frac{1}{a}}\right|$$

where r is the radius at which the maneuver occurs. Orbit correction strategies involve impulsive changes at perigee or apogee.

Polar Orbit Initial Mission Analysis

Polar orbits facilitate global earth observation by allowing a satellite to pass over both poles. These orbits often have inclinations i near 90 degrees. The orbital period for a circular orbit is derived similarly:

$$T = 2\pi\sqrt{\frac{a_{\text{polar}}^3}{\mu}}$$

The critical feature of a polar orbit is its high inclination, enabling maximal latitude coverage. Orbital elements must be tuned to ensure consecutive overpasses occur over progressively adjacent longitudinal tracks, exploiting the Earth's rotation beneath the orbit.

Sun-Synchronous Polar Orbits

A specialized type of polar orbit, the sun-synchronous orbit, achieves consistent illumination conditions over the targeted ground area. The theorem of nodal regression offers insight, with the rate of regression given by:

$$\dot{\Omega} = -\frac{3}{2} \frac{\sqrt{\mu} \, J_2 R_E^2}{a_{\text{polar}}^{7/2}} (1 - e^2)^{-2}$$

where R_E is the Earth's mean radius, and e the orbital eccentricity. The appropriate selection of a_{polar} and i ensures the orbit precesses synchronously with the motion of the sun.

Coverage Optimization in Polar Orbits

The design of polar orbital paths requires optimizing the swath width $\Delta\theta$ and revisit time. The swath width at nadir is given by:

$$\Delta\theta = 2 \arccos \left(\frac{R_E \cos(\text{inc})}{R_E + h} \right)$$

where h is the satellite's altitude. The analytical derivation of these coverage metrics demands rigorous examination of orbital parameters and satellite payload capabilities, ensuring mission objectives align with navigational efficiency and environmental constraints.

Configuration of Launch Elements for Polar Missions

The launch vehicle's trajectory must be accurately tailored to insert the satellite into the desired polar orbit. Launch azimuths set the inclination, while energy optimization focuses on propellant-efficient transfer stages. The optimal insertion angle γ maximizes payload mass m_p given escape velocity constraints:

$$V_{\text{esc}} = \sqrt{\frac{2\mu}{R_E + h}}$$

where energy conservation principles guide the ascent phase planning and motor burn profiles for achieving specified orbital altitudes without fiscal overspend.

Propellant Budgeting for Orbital Adjustments in Polar Regions

Adjustments within polar orbits often involve delta-v budgeting aligned with mission-specific observational requirements. Transition maneuvers between sun-synchronous and true polar inclinations or altitude adjustments are guided by propellant mass efficiency codes such as:

$$\Delta m = m_f \left(1 - \exp\left(-\frac{\Delta V}{I_{\text{sp}} g_0}\right)\right)$$

where I_{sp} is the specific impulse, and g_0 the standard gravity. These calculations dictate allowable deviations from nominal orbital paths, fundamental for responsive mission adaptations.

Python Code Snippet

Below is a Python code snippet encompassing the core computational elements of geosynchronous and polar orbit calculations including the orbital period, station-keeping maneuvers, and polar orbit characteristics.

```
import numpy as np

def geosynchronous_orbit_semi_major_axis(T, mu=3.986e14):
    '''
    Calculate the semi-major axis for a geosynchronous orbit.
    :param T: Orbital period in seconds (86,400 seconds for
    ↪ geosynchronous).
    :param mu: Earth's standard gravitational parameter.
    :return: Semi-major axis in meters.
    '''
    return (mu * T**2 / (4 * np.pi**2))**(1/3)

def radial_distance_change(mu, r, a):
    '''
    Calculate the radial distance change for station-keeping.
    :param mu: Earth's standard gravitational parameter.
    :param r: Current radial distance.
    :param a: Semi-major axis.
    :return: Delta V required for station-keeping.
    '''
    return np.sqrt(mu) * abs(np.sqrt(2 / r) - np.sqrt(1 / a))

def polar_orbit_period(a_polar, mu=3.986e14):
```

```python
    '''
    Calculate the orbital period for a polar orbit.
    :param a_polar: Semi-major axis for polar orbit.
    :param mu: Earth's standard gravitational parameter.
    :return: Orbital period in seconds.
    '''
    return 2 * np.pi * np.sqrt(a_polar**3 / mu)

def sun_synchronous_inclination(mu, J2, R_E, a_polar, e=0):
    '''
    Compute the inclination for a sun-synchronous polar orbit.
    :param mu: Earth's gravitational parameter.
    :param J2: Oblateness factor.
    :param R_E: Earth's radius.
    :param a_polar: Semi-major axis for polar orbit.
    :param e: Eccentricity of the orbit.
    :return: Inclination in degrees.
    '''
    factor = (3/2) * (np.sqrt(mu) * J2 * R_E**2) / ((a_polar**(7/2))
       ↪  * (1 - e**2)**2)
    i_rad = np.arccos(-factor)
    return np.degrees(i_rad)

def revisit_time(R_E, inc, h):
    '''
    Calculate swath width for a polar orbit.
    :param R_E: Earth's radius.
    :param inc: Inclination.
    :param h: Altitude.
    :return: Swath width in radians.
    '''
    return 2 * np.arccos((R_E * np.cos(np.radians(inc))) / (R_E +
       ↪  h))

# Example geosynchronous orbit calculation
T_geosync = 86400  # seconds
a_geo = geosynchronous_orbit_semi_major_axis(T_geosync)
print("Geosynchronous orbit semi-major axis:", a_geo)

# Example station-keeping calculation
r = 42164e3  # example radius in meters
Delta_V = radial_distance_change(3.986e14, r, a_geo)
print("Station-keeping Delta V:", Delta_V)

# Example polar orbit and sun-synchronous calculation
a_polar = 7000e3  # example semi-major axis in meters
T_polar = polar_orbit_period(a_polar)
print("Polar orbit period:", T_polar)

i = sun_synchronous_inclination(3.986e14, 1.08263e-3, 6371e3,
   ↪  a_polar)
print("Sun-synchronous inclination:", i)
```

```
# Example revisit time
swath_width = revisit_time(6371e3, 98.6, 700e3)
print("Swath width:", swath_width)
```

This code defines several key functions necessary for the computation of orbit elements:

- `geosynchronous_orbit_semi_major_axis` function computes the semi-major axis needed to establish a geosynchronous orbit.

- `radial_distance_change` calculates the Delta V required for station-keeping maneuvers in orbit.

- `polar_orbit_period` calculates the orbital period for a given polar orbit semi-major axis.

- `sun_synchronous_inclination` computes the orbital inclination necessary for maintaining a sun-synchronous orbit.

- `revisit_time` calculates the ground track swath width of a polar orbit, accounting for inclination and altitude.

The final block of code provides examples of computing these elements using hypothetical data inputs.

Chapter 34

Interplanetary Transfer Calculations

Foundations of Interplanetary Mechanics

Interplanetary transfer calculations are pivotal in the successful execution of missions between celestial bodies. These computations are fundamentally reliant on the principles of celestial mechanics, where the interplay of gravitational influences from multiple celestial entities dictates trajectory planning. The gravitational interactions are quantified using Kepler's laws and Newtonian mechanics.

Hohmann Transfer Calculations

The Hohmann transfer orbit remains a prominent strategy for interplanetary travel due to its energy efficiency. It is defined as an elliptical orbit that touches both the orbit of the departure planet and the destination planet, requiring two primary impulsive burns. For a transfer between two circular coplanar orbits, the semi-major axis a_H of the Hohmann transfer is given by:

$$a_H = \frac{r_1 + r_2}{2}$$

where r_1 and r_2 denote the radii of the inner and outer orbits, respectively. The total velocity change ΔV required for this maneuver is the sum of the velocity changes at periapsis and apoapsis:

$$\Delta V = \Delta V_1 + \Delta V_2$$

where:

$$\Delta V_1 = \sqrt{\frac{\mu}{r_1}} \left(\sqrt{\frac{2r_2}{r_1 + r_2}} - 1 \right)$$

$$\Delta V_2 = \sqrt{\frac{\mu}{r_2}} \left(1 - \sqrt{\frac{2r_1}{r_1 + r_2}} \right)$$

Bi-Elliptic Transfer Considerations

For specific situations, a bi-elliptic transfer becomes superior in terms of energy requirements compared to a Hohmann transfer, particularly when the desired orbit change involves large semi-major axes. The bi-elliptic transfer involves two intermediary ellipses and requires three impulse burns. The total ΔV is now expressed as:

$$\Delta V = \Delta V_1 + \Delta V_2 + \Delta V_3$$

where:

$$\Delta V_1 = \sqrt{\frac{\mu}{r_1}} \left(\sqrt{\frac{2r_b}{r_1 + r_b}} - 1 \right)$$

$$\Delta V_2 = \sqrt{\frac{\mu}{r_b}} \left(\sqrt{\frac{2r_2}{r_2 + r_b}} - \sqrt{\frac{2r_1}{r_1 + r_b}} \right)$$

$$\Delta V_3 = \sqrt{\frac{\mu}{r_2}} \left(1 - \sqrt{\frac{2r_b}{r_2 + r_b}} \right)$$

and r_b is the intermediate apoapsis radius.

Influence of Gravitational Assists

Gravitational assists or slingshot maneuvers exploit the relative motion and gravitational fields of larger celestial bodies to alter a spacecraft's trajectory and speed without using on-board propulsion. The velocity boost gained through such maneuver can be approximated by the vis-viva equation modified for flybys:

$$V_{\text{after}} = \sqrt{V_{\text{before}}^2 + 2V_P V_\infty}$$

where V_P is the velocity of the planetary body relative to the sun, and V_∞ is the excess velocity at infinity.

Patched Conic Approximations

For complex interplanetary missions, where multiple gravitational fields influence the trajectory, patched conic approximations simplify transfer calculations. This model divides the trajectory into conic segments, each governed by the dominant gravitational influence, seamlessly patched at the boundaries. The transition between the spheres of influence is managed by ensuring the continuity of the trajectory at the patch points. Calculations are performed separately for each segment, typically using standard orbital mechanics equations, and then combined to derive the complete transfer path.

N-Body Problem and Numerical Methods

Inherent challenges arise due to the intricacies of multi-body gravitational interactions in the solar system, traditionally captured by the N-body problem. Direct analytical solutions being infeasible, reliance on numerical simulations is quintessential. Numerical integration methods, such as Runge-Kutta or symplectic integrators, provide precise trajectory predictions under varying gravitational influences by iteratively solving the equations of motion.

The governing equations for any spacecraft subjected to N gravitational accelerating bodies can be described by:

$$\frac{d^2\mathbf{r}}{dt^2} = -\sum_{i=1}^{N} \frac{\mu_i(\mathbf{r} - \mathbf{r}_i)}{\|\mathbf{r} - \mathbf{r}_i\|^3}$$

where $\mathbf{r}$ is the spacecraft's position vector, μ_i is the gravitational parameter of the ith body, and $\mathbf{r}_i$ represents its position vector relative to an inertial reference frame.

Optimization of Trajectory Pathways

Interplanetary trajectories must be optimized to minimize propellant consumption and mission duration, balancing gravitational assists, transfer orbits, and external perturbations. Optimization algorithms, such as genetic algorithms and particle swarm methods, are deployed to refine path variables. The objective function typically involves minimizing ΔV while adhering to mission constraints such as time of flight, arrival hyperbolic excess speed, and spacecraft systems limitations.

By embracing these computational models and calculations, the rigorous demands of interplanetary travel are met, leveraging both deterministic physics and dynamic optimization strategies to secure mission successes across the cosmos.

Python Code Snippet

Below is a Python code snippet that encompasses the core computational elements needed for interplanetary transfer calculations, including Hohmann transfer orbit, bi-elliptic transfer considerations, gravitational assists, and more.

```python
import numpy as np
from scipy.integrate import solve_ivp

def hohmann_transfer_delta_v(r1, r2, mu):
    '''
    Calculate delta-v for a Hohmann transfer orbit.
    :param r1: Radius of the initial orbit.
    :param r2: Radius of the final orbit.
    :param mu: Standard gravitational parameter (GM).
    :return: Total delta-v required.
    '''
    a_H = (r1 + r2) / 2
    delta_v1 = np.sqrt(mu / r1) * (np.sqrt(2 * r2 / (r1 + r2)) - 1)
    delta_v2 = np.sqrt(mu / r2) * (1 - np.sqrt(2 * r1 / (r1 + r2)))
    return delta_v1 + delta_v2

def bi_elliptic_transfer_delta_v(r1, rb, r2, mu):
    '''
    Calculate delta-v for a bi-elliptic transfer.
    :param r1: Radius of the initial orbit.
    :param rb: Radius of the intermediate orbit.
    :param r2: Radius of the final orbit.
    :param mu: Standard gravitational parameter (GM).
    :return: Total delta-v required.
```

```python
    '''
    delta_v1 = np.sqrt(mu / r1) * (np.sqrt(2 * rb / (r1 + rb)) - 1)
    delta_v2 = np.sqrt(mu / rb) * (np.sqrt(2 * r2 / (r2 + rb)) -
       np.sqrt(2 * r1 / (r1 + rb)))
    delta_v3 = np.sqrt(mu / r2) * (1 - np.sqrt(2 * rb / (r2 + rb)))
    return delta_v1 + delta_v2 + delta_v3

def gravitational_assist(v_before, v_p, v_inf):
    '''
    Calculate the velocity after a gravitational assist.
    :param v_before: Velocity before the flyby.
    :param v_p: Velocity of the planetary body.
    :param v_inf: Excess velocity at infinity.
    :return: Velocity after the flyby.
    '''
    return np.sqrt(v_before**2 + 2 * v_p * v_inf)

def n_body_simulation(initial_conditions, mu_values, t_span,
   num_bodies=3):
    '''
    Perform an N-body simulation using numerical integration.
    :param initial_conditions: Initial state vectors of the bodies.
    :param mu_values: Gravitational parameters of the bodies.
    :param t_span: Time span for the simulation.
    :param num_bodies: Number of bodies involved in the simulation.
    :return: Results of the simulation.
    '''
    def equations_of_motion(t, y):
        positions = y[:num_bodies * 3].reshape((num_bodies, 3))
        velocities = y[num_bodies * 3:].reshape((num_bodies, 3))
        accelerations = np.zeros((num_bodies, 3))
        for i in range(num_bodies):
            for j in range(num_bodies):
                if i != j:
                    r_vec = positions[j] - positions[i]
                    r_mag = np.linalg.norm(r_vec)
                    accelerations[i] += mu_values[j] * r_vec /
                       r_mag**3
        return np.concatenate((velocities.flatten(),
           accelerations.flatten()))

    return solve_ivp(equations_of_motion, t_span,
       initial_conditions, method='RK45')

# Example calculations using the defined functions
mu_earth = 3.986e14   # Gravitational parameter for Earth in m^3/s^2
r_earth = 6.371e6   # Earth's radius in meters
r1 = r_earth + 500e3   # Initial orbit: 500 km above Earth's surface
r2 = r_earth + 36e6   # Geostationary orbit
r_b = 150e6           # Intermediate orbit for bi-elliptic transfer

delta_v_hohmann = hohmann_transfer_delta_v(r1, r2, mu_earth)
```

```
delta_v_bi_elliptic = bi_elliptic_transfer_delta_v(r1, r_b, r2,
↪    mu_earth)

v_before = 7.8e3   # Example velocity before flyby in m/s
v_p = 30e3         # Planetary velocity in m/s
v_inf = 3e3        # Excess velocity at infinity in m/s
v_after = gravitational_assist(v_before, v_p, v_inf)

# Output results
print("Hohmann Transfer V:", delta_v_hohmann)
print("Bi-Elliptic Transfer V:", delta_v_bi_elliptic)
print("Velocity after Gravitational Assist:", v_after)
```

This code defines several key functions necessary for the implementation of interplanetary transfer calculations:

- `hohmann_transfer_delta_v` function computes the delta-v required for a Hohmann transfer orbit for efficient interplanetary missions.

- `bi_elliptic_transfer_delta_v` calculates the delta-v for a bi-elliptic transfer, advantageous for certain orbit changes.

- `gravitational_assist` estimates the velocity change achieved through slingshot maneuvers around celestial bodies.

- `n_body_simulation` performs an N-body numerical simulation to predict gravitational interactions in space.

The final block of code provides examples of computing these elements with dummy data, helpful for understanding interplanetary transfer strategies.

Chapter 35

Ion Propulsion System Calculations

Fundamentals of Ion Propulsion

Ion propulsion is distinguished by its utilization of electricity to ionize a propellant, generating ions that are accelerated by electric or magnetic fields to produce thrust. The efficiency of these systems is predominantly quantified by their specific impulse (I_{sp}), which often surpasses that of conventional chemical propulsion. The fundamental equation for the electric thrust (F) generated is expressed as:

$$F = \dot{m} v_e \quad (35.1)$$

where $\dot{m}$ is the mass flow rate of the ionized propellant and v_e is the effective exhaust velocity. For ion thrusters, the effective I_{sp} is given by:

$$I_{sp} = \frac{v_e}{g_0} \quad (35.2)$$

with g_0 representing standard gravitational acceleration.

Ionization and Acceleration of Propellant

The ionization process in ion propulsion is a critical aspect, where electrons are freed from atoms, producing positively charged ions. The energy required for this ionization (E_i) can be calculated by:

$$E_i = \frac{e \cdot U_{\text{ion}} \cdot N_i}{q} \quad (35.3)$$

Here, e denotes the elementary charge, U_{ion} is the ionization potential, N_i is the number of ions produced, and q is the charge per ion.

Once ionized, the ions are subjected to electric fields created by grids, resulting in their acceleration. The potential difference (V) across the grids dictates the final velocity (v_e) of the ions:

$$v_e = \sqrt{\frac{2eV}{m_i}} \quad (35.4)$$

where m_i is the ion mass.

Thrust Power Calculations

The power (P) required to sustain ion propulsion is a function of both the thrust produced and the efficiency of ionization and acceleration. It can be deduced by:

$$P = \frac{\dot{m} v_e^2}{2\eta} \quad (35.5)$$

η symbolizes the system efficiency, acknowledging losses in electric power conversion and ion production.

Grid Erosion and System Longevity

Ion propulsion system longevity is contingent upon grid erosion, fundamentally governed by sputtering yields (Y):

$$Y = \frac{\text{mass of target material sputtered}}{\text{mass of incident ions}} \quad (35.6)$$

Minimizing erosion rates ensures extended operational capability, crucial for deep space missions. The effect of erosion on thrust can be quantified by:

$$F_{\text{reduced}} = F_0 - K \cdot \Delta L \quad (35.7)$$

where F_0 is the initial thrust, K is a material constant, and ΔL is the material loss due to sputtering.

Optimization Techniques for Deep Space Missions

The optimization of ion propulsion systems for deep space missions involves a comprehensive understanding of propellant utilization and power availability. The thrust-to-power ratio (T/P) is a fundamental metric for evaluating system performance:

$$\frac{T}{P} = \frac{2\eta}{v_e} \tag{35.8}$$

Potential deep space propulsion system designs employ optimization algorithms such as genetic algorithms and machine learning techniques to maximize T/P while adhering to mission power constraints.

Incorporating real-time feedback loops allows dynamic adjustment of operational parameters to achieve desired trajectory modifications with minimal energy expenditure. Advanced computational simulations play a pivotal role in parameter space exploration, ensuring the thruster operates at peak performance across varying mission profiles.

Mathematical Modeling in Ion Thruster Design

The mathematical modeling of ion propulsion systems necessitates solving complex differential equations governing ion trajectories and grid fields. Utilizing techniques such as finite element analysis, the electric field (E) within the accelerator can be computed, leading to the optimization of grid geometry:

$$E = -\nabla V \tag{35.9}$$

By coupling these field equations with ion transport equations, an amplified understanding of space charge effects, plume divergence, and potential recombination phenomena is realized, enabling precise thruster blueprinting for extended voyages in the cosmos.

Python Code Snippet

Below is a Python code snippet that demonstrates the essential calculations and algorithms related to ion propulsion systems, including thrust calculations, ionization energy, power requirements, optimization techniques, and insights into grid erosion and system longevity.

```python
import numpy as np

def thrust_ion_propulsion(mass_flow_rate, velocity_exhaust):
    '''
    Calculate the thrust produced by ion propulsion.
    :param mass_flow_rate: Mass flow rate of ionized propellant
    ↪ (kg/s).
    :param velocity_exhaust: Effective exhaust velocity (m/s).
    :return: Thrust (N).
    '''
    return mass_flow_rate * velocity_exhaust

def specific_impulse(velocity_exhaust, g0=9.81):
    '''
    Calculate the specific impulse of the propulsion system.
    :param velocity_exhaust: Effective exhaust velocity (m/s).
    :param g0: Standard gravitational acceleration (m/s^2).
    :return: Specific impulse (s).
    '''
    return velocity_exhaust / g0

def ionization_energy(elementary_charge, ionization_potential,
    ↪ num_ions, charge_per_ion):
    '''
    Calculate the energy required for ionization.
    :param elementary_charge: Charge of an electron (Coulombs).
    :param ionization_potential: Ionization potential (Volts).
    :param num_ions: Number of ions produced.
    :param charge_per_ion: Charge per ion.
    :return: Ionization energy (Joules).
    '''
    return elementary_charge * ionization_potential * num_ions /
    ↪ charge_per_ion

def ion_velocity(exhaust_potential, ion_mass,
    ↪ elementary_charge=1.602e-19):
    '''
    Calculate the velocity of ions accelerated by potential
    ↪ difference.
    :param exhaust_potential: Potential difference across the grids
    ↪ (Volts).
    :param ion_mass: Mass of the ion (kg).
    :param elementary_charge: Charge of an electron (Coulombs).
```

```python
    :return: Ion velocity (m/s).
    '''
    return np.sqrt(2 * elementary_charge * exhaust_potential /
    ↪  ion_mass)

def thrust_power(mass_flow_rate, velocity_exhaust, efficiency):
    '''
    Calculate the power required for ion propulsion.
    :param mass_flow_rate: Mass flow rate of ionized propellant
    ↪  (kg/s).
    :param velocity_exhaust: Effective exhaust velocity (m/s).
    :param efficiency: System efficiency.
    :return: Power required (Watts).
    '''
    return (mass_flow_rate * velocity_exhaust**2) / (2 * efficiency)

def grid_erosion_sputtering(sputtered_mass, incident_ion_mass):
    '''
    Calculate the sputtering yield for grid erosion.
    :param sputtered_mass: Mass of target material sputtered (kg).
    :param incident_ion_mass: Mass of incident ions (kg).
    :return: Sputtering yield.
    '''
    return sputtered_mass / incident_ion_mass

def reduced_thrust(initial_thrust, material_constant,
↪  material_loss):
    '''
    Calculate the reduced thrust due to grid erosion.
    :param initial_thrust: Initial thrust (N).
    :param material_constant: Material-specific constant.
    :param material_loss: Material loss due to sputtering (kg).
    :return: Reduced thrust (N).
    '''
    return initial_thrust - material_constant * material_loss

def thrust_to_power_ratio(thrust, power):
    '''
    Calculate the thrust-to-power ratio of propulsion system.
    :param thrust: Produced thrust (N).
    :param power: Power used (Watts).
    :return: Thrust-to-power ratio (N/W).
    '''
    return thrust / power

# Practical example with actual inputs
mass_flow_rate = 0.0002   # kg/s
velocity_exhaust = ion_velocity(3000, 2.18e-25)   # m/s
efficiency = 0.7

# Demonstrating calculations
thrust = thrust_ion_propulsion(mass_flow_rate, velocity_exhaust)
isp = specific_impulse(velocity_exhaust)
```

```
power = thrust_power(mass_flow_rate, velocity_exhaust, efficiency)

print("Thrust:", thrust, "N")
print("Specific Impulse:", isp, "s")
print("Power Required:", power, "W")
```

This code snippet provides a comprehensive set of functions necessary for modeling the ion propulsion system:

- `thrust_ion_propulsion` function calculates the thrust produced by the ionized propellants.

- `specific_impulse` computes the specific impulse, a measure of propulsion efficiency.

- `ionization_energy` determines the energy needed to ionize the propellant.

- `ion_velocity` calculates the velocity of ions based on the potential difference and ion mass.

- `thrust_power` evaluates the power required for a given mass flow rate and velocity.

- `grid_erosion_sputtering` models the sputtering-induced erosion of grids.

- `reduced_thrust` tracks the impact of material erosion on thrust.

- `thrust_to_power_ratio` assesses the efficiency of the propulsion system with respect to power usage.

An example calculation is performed with a hypothetical mass flow rate and efficiency settings, showcasing application of the fundamental equations in real-world scenarios.

Chapter 36

Hybrid Rocket Propulsion

Introduction to Hybrid Rocket Propulsion

Hybrid rocket propulsion systems represent a unique synthesis of solid and liquid propulsion technologies. They utilize a solid fuel and a liquid or gaseous oxidizer, combining the advantages of both systems to achieve flexible throttling, inherent safety, and simplicity in design. The fundamental thrust generation mechanism involves the combustion of solid fuel with a liquid oxidizer, producing expansive gases that are accelerated through a rocket nozzle.

Combustion Processes in Hybrid Rockets

The combustion chamber of a hybrid rocket is where the propellant's solid phase reacts with the injected oxidizer. This process is characterized by complex thermochemical interactions and can be represented by the equation:

$$\text{Fuel (solid)} + \text{Oxidizer (liquid)} \rightarrow \text{Products} + \text{Energy} \quad (36.1)$$

The efficiency of this reaction is reliant on the oxidizer-to-fuel ratio (O/F), a critical parameter influencing the flame temperature and resultant exhaust characteristics. This ratio is calculated as:

$$\text{O/F} = \frac{\dot{m}_{\text{oxidizer}}}{\dot{m}_{\text{fuel}}}$$

where $\dot{m}_{\text{oxidizer}}$ is the oxidizer mass flow rate, and $\dot{m}_{\text{fuel}}$ is the fuel regression rate treated as a mass flow.

Solid Fuel Regression Rates

In hybrid rockets, the fuel regression rate (r) is a crucial parameter defining the fuel consumption rate per unit surface area and is typically modeled empirically as:

$$r = a \cdot G^n \tag{36.2}$$

where a is a regression rate coefficient and n is the regression rate exponent. The oxidizer mass flux (G) is expressed as:

$$G = \frac{\dot{m}_{\text{oxidizer}}}{A_{\text{port}}}$$

Here, A_{port} refers to the port cross-sectional area through which the oxidizer flows. The regression rate determines the evolution of the port diameter and thus impacts the hybrid motor's overall thrust profile.

Thermodynamic Analysis

The performance of a hybrid rocket, much like other propulsion systems, can be assessed by evaluating the specific impulse (I_{sp}), which defines the efficiency of the exhaust exit velocity v_e under specific gravitational conditions. Expressed as:

$$I_{sp} = \frac{v_e}{g_0} \tag{36.3}$$

The exhaust velocity (v_e) is derived from the combustion chamber properties and the nozzle expansion ratio, with the relationship given by:

$$v_e = \sqrt{\frac{2\gamma R T_c}{\gamma - 1}\left(1 - \left(\frac{p_e}{p_c}\right)^{\frac{\gamma-1}{\gamma}}\right)}$$

Here, γ is the specific heat ratio, R is the gas constant, T_c the chamber temperature, p_e the exit pressure, and p_c the chamber pressure.

Design Considerations for Hybrid Systems

The architecture of a hybrid rocket requires careful attention to the material properties of the solid fuel, the reactivity of the oxidizer, and the structural integrity under pressure conditions. The relationship between the chamber pressure (p_c) and the nozzle thrust can be encapsulated by:

$$F = (\dot{m}v_e) + (p_e - p_a)A_e \tag{36.4}$$

where F is the thrust, $\dot{m}$ the total mass flow rate, p_a the ambient pressure, and A_e the exit area.

The combustor design often involves the evaluation of heat transfer coefficients, combustion efficiency, and potential erosive burning impact on the solid fuel. Accurate thermophysical modeling and real-time monitoring of these parameters enhance the predictive capabilities of hybrid rocket performance.

Hybrid Rocket Propellant Materials

Material selection for hybrid rocket propellants involves a balance between combustion characteristics, mechanical properties, and energetic yield. Many modern systems employ hydroxyl-terminated polybutadiene (HTPB) as a fuel matrix due to its favorable mechanical stability and energy output.

$$\Delta H_{\text{comb}} = \int C_p \, dT$$

This equation calculates the total heat of combustion through the integral of the specific heat capacity (C_p) over the temperature range experienced during burn.

Material testing under actual operating conditions is essential to validate the theoretical predictions and ensure consistent performance across applications.

System Integration Principles

For efficient system integration, synchronization of pressurization and fuel consumption rates in hybrid motors must be achieved. This involves utilizing precise regulatory mechanisms for oxidizer flow and incorporating feedback loops to modulate performance in response to dynamic environmental conditions.

Detailed design optimization, such as variable geometry nozzles and novel injector systems, aids in maintaining optimal internal flow characteristics, which are vital for performance maximization in hybrid rocket engines.

Python Code Snippet

Below is a Python code snippet that encompasses the core computational elements described in this chapter, including calculations for oxidizer-to-fuel ratio, solid fuel regression rate, specific impulse, and more.

```python
import numpy as np

def oxidizer_to_fuel_ratio(m_oxidizer, m_fuel):
    '''
    Calculate the oxidizer to fuel ratio (O/F).
    :param m_oxidizer: Mass flow rate of the oxidizer.
    :param m_fuel: Mass flow rate of the fuel.
    :return: O/F ratio.
    '''
    return m_oxidizer / m_fuel

def regression_rate(a, G, n):
    '''
    Calculate the solid fuel regression rate in hybrid rockets.
    :param a: Regression rate coefficient.
    :param G: Oxidizer mass flux.
    :param n: Regression rate exponent.
    :return: Fuel regression rate.
    '''
    return a * (G ** n)

def oxidizer_mass_flux(m_oxidizer, A_port):
    '''
    Calculate the oxidizer mass flux.
    :param m_oxidizer: Mass flow rate of the oxidizer.
    :param A_port: Port cross-sectional area.
    :return: Oxidizer mass flux.
    '''
```

```python
    return m_oxidizer / A_port

def specific_impulse(v_e, g_0):
    '''
    Calculate the specific impulse (Isp) of a rocket.
    :param v_e: Effective exhaust velocity.
    :param g_0: Standard gravity.
    :return: Specific impulse.
    '''
    return v_e / g_0

def exhaust_velocity(gamma, R, T_c, p_e, p_c):
    '''
    Calculate the exhaust velocity of a rocket.
    :param gamma: Specific heat ratio.
    :param R: Gas constant.
    :param T_c: Chamber temperature.
    :param p_e: Exit pressure.
    :param p_c: Chamber pressure.
    :return: Exhaust velocity.
    '''
    term1 = 2 * gamma * R * T_c / (gamma - 1)
    term2 = 1 - (p_e / p_c) ** ((gamma - 1) / gamma)
    return np.sqrt(term1 * term2)

def thrust(m_dot, v_e, p_e, p_a, A_e):
    '''
    Calculate the thrust produced by a rocket.
    :param m_dot: Total mass flow rate.
    :param v_e: Exhaust velocity.
    :param p_e: Exit pressure.
    :param p_a: Ambient pressure.
    :param A_e: Exit area.
    :return: Thrust.
    '''
    return m_dot * v_e + (p_e - p_a) * A_e

# Example parameters
m_oxidizer = 5.0    # kg/s
m_fuel = 1.0    # kg/s
a = 0.1
G = 250    # kg/m^2/s
n = 0.8
A_port = 0.05    # m^2
gamma = 1.2
R = 287    # J/kg-K
T_c = 3500    # Kelvin
p_e = 101325    # N/m^2
p_c = 4000000    # N/m^2
g_0 = 9.81    # m/s^2
A_e = 0.1    # m^2
p_a = 101000    # N/m^2
```

```
# Perform calculations
of_ratio = oxidizer_to_fuel_ratio(m_oxidizer, m_fuel)
fuel_regression_rate = regression_rate(a, G, n)
ox_flux = oxidizer_mass_flux(m_oxidizer, A_port)
v_e = exhaust_velocity(gamma, R, T_c, p_e, p_c)
Isp = specific_impulse(v_e, g_0)
thrust_value = thrust(m_oxidizer + m_fuel, v_e, p_e, p_a, A_e)

print("O/F Ratio:", of_ratio)
print("Fuel Regression Rate:", fuel_regression_rate)
print("Oxidizer Mass Flux:", ox_flux)
print("Exhaust Velocity:", v_e)
print("Specific Impulse (Isp):", Isp)
print("Thrust:", thrust_value)
```

This code defines several essential functions for calculating the key parameters involved in hybrid rocket propulsion:

- `oxidizer_to_fuel_ratio` function computes the oxidizer-to-fuel ratio, crucial for combustion efficiency.

- `regression_rate` calculates the fuel regression rate, impacting the thrust profile.

- `oxidizer_mass_flux` is used to find the oxidizer mass flux in the rocket.

- `specific_impulse` gives the specific impulse of the rocket, a measure of efficiency.

- `exhaust_velocity` computes the exhaust velocity, which feeds into the specific impulse calculation.

- `thrust` calculates the total thrust produced by the rocket.

The final section of the code provides examples of these computations using example data for a typical hybrid rocket propulsion scenario.

Chapter 37

Rocket Propellant Thermochemistry

Fundamental Thermochemistry of Rocket Propellants

The study of rocket propellant thermochemistry entails understanding the complex interactions occurring during combustion, which lead to heat release and energy conversion. The energy released by propellants is dictated by their chemical composition and is an essential factor in determining propulsion efficiency. The enthalpy of formation (ΔH_f) of the reactants and products is crucial for calculating the overall heat release during combustion, with the reaction energy represented as:

$$\Delta H_{\text{reaction}} = \sum \Delta H_f(\text{products}) - \sum \Delta H_f(\text{reactants}) \quad (37.1)$$

The thermodynamic assessment involves evaluating the combustion enthalpy, assessing factors such as pressure, temperature, and reaction kinetics.

Heat of Combustion Calculations

The heat of combustion for a given propellant composition determines the potential energy convertible into kinetic energy for propulsion purposes. The calculation employs the specific heat capacity (C_p) over the temperature range:

$$\Delta H_{\text{comb}} = \int_{T_1}^{T_2} C_p \, dT \tag{37.2}$$

This integral evaluates the thermal energy difference as the system transitions from an initial state T_1 to a final state T_2.

Chemical Equilibrium and Thermodynamic Efficiency

Chemical equilibrium within the combustion chamber of a rocket engine is achieved when the rates of the forward and reverse reactions equalize, yielding maximum energy efficiency. The equilibrium constant (K) for these reactions can be calculated using the Gibbs free energy (ΔG) relationship:

$$K = e^{-\Delta G/(RT)} \tag{37.3}$$

where R is the universal gas constant and T is the absolute temperature. The equilibrium composition influences the specific impulse (I_{sp}) and overall propulsion effectiveness.

Adiabatic Flame Temperature

The adiabatic flame temperature represents the theoretical maximum temperature achievable under constant pressure and assuming no heat loss to the surroundings. It is a determinant of the energy conversion potential of propellants, estimated using:

$$T_{\text{ad}} = T_0 + \frac{\Delta H_{\text{comb}}}{\sum n_i C_{p,i}} \tag{37.4}$$

where T_0 is the initial temperature, n_i represents the moles of each product, and $C_{p,i}$ is their respective specific heat capacity.

Thermochemical Reaction Kinetics

The kinetics of the reaction process in rocket propellants influence the rate of energy release and the combustion efficiency. The rate of reaction is modeled using the Arrhenius equation:

$$k = Ae^{-E_a/(RT)} \qquad (37.5)$$

Here, k signifies the rate constant, A is the pre-exponential factor, E_a denotes the activation energy, and T is the temperature. Reaction rate control is vital for maintaining stable combustion within the chamber.

Energy Efficiency Metrics

The energy efficiency of a propellant can be qualitatively described by metrics which relate heat release to propulsion performance. The specific impulse, which links directly to the effective exhaust velocity (v_e) and gravitational influence, is calculated with the expression:

$$I_{sp} = \frac{v_e}{g_0} \qquad (37.6)$$

In conjunction with this, the energy density and the mass flow parameters are used to derive efficiencies. An optimized thermochemical profile ensures that the theoretical specific impulse approaches real-world outcomes, giving rise to effective propulsion systems.

These calculations and frameworks form the foundation of Rocket Propellant Thermochemistry, providing key insights into the design and application of energy-efficient propulsion mechanisms in aerospace engineering.

Python Code Snippet

Below is a Python code snippet that encompasses the core computational elements for thermochemical calculations in rocket propulsion, including heat of reaction, heat combustion, chemical equilibrium, adiabatic flame temperature, reaction kinetics, and energy efficiency metrics.

```
import numpy as np

def calculate_heat_of_reaction(delta_H_f_products,
    delta_H_f_reactants):
    """
    Calculate the heat of reaction based on the enthalpy of
        formation
```

```
    :param delta_H_f_products: List of enthalpy of formation for
    ↪    products
    :param delta_H_f_reactants: List of enthalpy of formation for
    ↪    reactants
    :return: Heat of reaction
    """
    return sum(delta_H_f_products) - sum(delta_H_f_reactants)

def calculate_heat_of_combustion(C_p, T1, T2):
    """
    Calculate the heat of combustion over a temperature range using
    ↪    specific heat capacity
    :param C_p: Specific heat capacity
    :param T1: Initial temperature
    :param T2: Final temperature
    :return: Heat of combustion
    """
    return np.trapz([C_p(t) for t in np.linspace(T1, T2, 100)],
    ↪    dx=(T2-T1)/100)

def calculate_equilibrium_constant(delta_G, T):
    """
    Calculate equilibrium constant from Gibbs free energy
    :param delta_G: Change in Gibbs free energy
    :param T: Temperature
    :return: Equilibrium constant
    """
    R = 8.314  # Universal gas constant J/(mol*K)
    return np.exp(-delta_G / (R * T))

def calculate_adiabatic_flame_temperature(T0, delta_H_comb, n_Cp):
    """
    Calculate the adiabatic flame temperature
    :param T0: Initial temperature
    :param delta_H_comb: Heat of combustion
    :param n_Cp: Array of moles multiplied by specific heat
    ↪    capacities of products
    :return: Adiabatic flame temperature
    """
    return T0 + delta_H_comb / sum(n_Cp)

def calculate_reaction_rate_constant(A, E_a, T):
    """
    Calculate the reaction rate constant using the Arrhenius
    ↪    equation
    :param A: Pre-exponential factor
    :param E_a: Activation energy
    :param T: Temperature
    :return: Reaction rate constant
    """
    R = 8.314  # Universal gas constant J/(mol*K)
    return A * np.exp(-E_a / (R * T))
```

```python
def calculate_specific_impulse(ve, g0=9.81):
    """
    Calculate the specific impulse
    :param ve: Effective exhaust velocity
    :param g0: Gravitational acceleration
    :return: Specific impulse
    """
    return ve / g0

# Example values
delta_H_f_products = [-393.5, -241.8]  # CO2 and H2O in kJ/mol
delta_H_f_reactants = [0, 0]  # O2 and H2 as elemental form in
↪  kJ/mol
C_p = lambda T: 29.1 + 0.01 * T  # Example specific heat capacity
↪  function in J/mol*K
T1, T2 = 298, 3500
delta_G = -237.1  # in kJ/mol
T = 298  # in Kelvin
A = 1.0e7
E_a = 75.0  # in kJ/mol
ve = 3000  # m/s

# Calculation Outputs
heat_of_reaction = calculate_heat_of_reaction(delta_H_f_products,
↪  delta_H_f_reactants)
heat_of_combustion = calculate_heat_of_combustion(C_p, T1, T2)
equilibrium_constant = calculate_equilibrium_constant(delta_G, T)
adiabatic_flame_temperature =
↪  calculate_adiabatic_flame_temperature(T1, heat_of_combustion,
↪  [1, 1])
reaction_rate_constant = calculate_reaction_rate_constant(A, E_a, T)
specific_impulse = calculate_specific_impulse(ve)

print("Heat of Reaction:", heat_of_reaction, "kJ/mol")
print("Heat of Combustion:", heat_of_combustion, "J/mol")
print("Equilibrium Constant:", equilibrium_constant)
print("Adiabatic Flame Temperature:", adiabatic_flame_temperature,
↪  "K")
print("Reaction Rate Constant:", reaction_rate_constant, "1/s")
print("Specific Impulse:", specific_impulse, "s")
```

This code defines several key functions necessary for the calculation of:

- calculate_heat_of_reaction computes the heat of reaction using the enthalpies of formation of reactants and products.

- calculate_heat_of_combustion calculates the heat of combustion over a temperature range using specific heat capacity.

- `calculate_equilibrium_constant` determines the equilibrium constant from Gibbs free energy changes.

- `calculate_adiabatic_flame_temperature` estimates the adiabatic flame temperature.

- `calculate_reaction_rate_constant` computes the reaction rate constant using the Arrhenius equation.

- `calculate_specific_impulse` calculates the specific impulse based on the effective exhaust velocity.

The final block of code provides examples of computing these elements using sample data for demonstration purposes.

Chapter 38

Combustion Efficiency Evaluation

Introduction to Combustion Efficiency

Combustion efficiency is a critical factor in the performance of rocket propulsion systems, reflecting the extent to which chemical energy is converted into useful kinetic energy. The efficiency of a combustion process can be evaluated by examining the completeness with which fuel reacts with an oxidizer and the thermodynamic conditions under which the reaction occurs. Mathematically, combustion efficiency (η_c) is defined as:

$$\eta_c = \frac{\text{Useful energy output}}{\text{Total energy input}} \tag{38.1}$$

Chemical Reaction Modeling

In rocket engines, the chemical reaction modeling provides insight into the combustion process and its efficiency. Stoichiometric calculations are fundamental in determining the ideal proportions of fuel and oxidizer required for complete combustion. Given a general combustion reaction:

$$\text{Fuel} + \text{Oxidizer} \rightarrow \text{Products} \tag{38.2}$$

The balancing of reactants ensures that the moles of reactants translate effectively into product moles. The stoichiometric coefficient (ν) allows for precise modeling, expressed as:

$$\sum \nu_i \text{Reactants}_i = \sum \nu_j \text{Products}_j \tag{38.3}$$

Heat Release and Energy Conversion

The energy conversion in combustion processes is quantified through the enthalpy changes (ΔH). The heat release per unit mass of fuel (q_{fuel}) is given by:

$$q_{\text{fuel}} = \frac{\Delta H_{\text{comb}}}{m_{\text{fuel}}} \tag{38.4}$$

where ΔH_{comb} is the heat of combustion, and m_{fuel} is the mass of the fuel. For effective conversion, maximizing q_{fuel} is imperative.

Efficiency Metrics and Performance

Various metrics are utilized to evaluate performance. The thermal efficiency (η_t) considers the conversion of heat energy into work and is defined by:

$$\eta_t = \frac{W_{\text{out}}}{Q_{\text{in}}} \tag{38.5}$$

where W_{out} is the work output and Q_{in} is the heat input.

Similarly, propulsion efficiency (η_p) is concerned with converting kinetic energy of exhaust gases into useful thrust:

$$\eta_p = \frac{2}{1 + \frac{c_f}{v_e}} \tag{38.6}$$

where c_f is the characteristic velocity and v_e is the effective exhaust velocity.

Combustion Kinetics

Combustion kinetics governs the rate at which chemical reactions occur, governed by factors such as temperature, pressure, and reactant concentration. The Arrhenius equation models the reaction rates (k):

$$k = A\exp\left(-\frac{E_a}{RT}\right) \qquad (38.7)$$

where A is the pre-exponential factor, E_a is the activation energy, R is the universal gas constant, and T is the temperature.

Thermochemical Efficiency Optimization

Optimizing thermochemical efficiency involves maximizing the energy conversion within the combustion process while minimizing losses. This entails the strategic selection of propellants with high specific energy and optimizing operational conditions—temperature, pressure ratio, and mixture ratio.

$$\eta_{\text{thermo}} = \frac{T_{\text{actual}} - T_{\text{ambient}}}{T_{\text{ideal}} - T_{\text{ambient}}} \qquad (38.8)$$

As a function of thermodynamic properties, the compression and expansion processes in the combustion chamber should be finely tuned to achieve peak efficiencies.

Through comprehensive mathematical analysis, the maximization of combustion efficiency can be pursued to yield superior propulsion performance, emphasizing efficient fuel utilization and optimal energy conversion parameters.

Python Code Snippet

Below is a Python code snippet that encompasses the essential computations related to evaluating combustion efficiency, including enthalpy changes, stoichiometric calculations, and performance metrics such as thermal and propulsion efficiency.

```
import numpy as np

def combustion_efficiency(useful_energy_output, total_energy_input):
    '''
    Calculate the combustion efficiency.
    :param useful_energy_output: Useful energy obtained from
    ↪    reaction.
    :param total_energy_input: Total energy provided by the fuel.
    :return: Combustion efficiency.
    '''
    return useful_energy_output / total_energy_input
```

```python
def stoichiometric_balance(reactants, products):
    '''
    Balance stoichiometric coefficients for a combustion reaction.
    :param reactants: Dictionary with reactants and their moles.
    :param products: Dictionary with products and their moles.
    :return: Boolean indicating if the reaction is balanced.
    '''
    reactant_sum = sum(reactants.values())
    product_sum = sum(products.values())
    return np.isclose(reactant_sum, product_sum)

def heat_release(enthalpy_combustion, mass_fuel):
    '''
    Calculate the heat release per unit mass of fuel.
    :param enthalpy_combustion: Heat of combustion.
    :param mass_fuel: Mass of the fuel.
    :return: Heat release per unit mass.
    '''
    return enthalpy_combustion / mass_fuel

def thermal_efficiency(work_output, heat_input):
    '''
    Calculate the thermal efficiency.
    :param work_output: Work obtained from the process.
    :param heat_input: Heat input to the system.
    :return: Thermal efficiency.
    '''
    return work_output / heat_input

def propulsion_efficiency(characteristic_velocity,
 ↪   exhaust_velocity):
    '''
    Calculate propulsion efficiency.
    :param characteristic_velocity: Characteristic velocity of the
    ↪   propellant.
    :param exhaust_velocity: Effective exhaust velocity.
    :return: Propulsion efficiency.
    '''
    return 2 / (1 + characteristic_velocity / exhaust_velocity)

def arrhenius_rate(A, E_a, T):
    '''
    Calculate the reaction rate using Arrhenius equation.
    :param A: Pre-exponential factor.
    :param E_a: Activation energy.
    :param T: Temperature in Kelvin.
    :return: Reaction rate.
    '''
    R = 8.314  # Universal gas constant in J/(mol*K)
    return A * np.exp(-E_a / (R * T))
```

```python
def thermochemical_efficiency(actual_temp, ambient_temp,
    ideal_temp):
    '''
    Calculate thermochemical efficiency.
    :param actual_temp: Actual operating temperature.
    :param ambient_temp: Ambient temperature.
    :param ideal_temp: Ideal thermodynamic temperature.
    :return: Thermochemical efficiency.
    '''
    return (actual_temp - ambient_temp) / (ideal_temp -
        ambient_temp)

# Example usage of functions for given parameters
useful_energy = 1000  # in Joules
total_energy = 1200   # in Joules
eta_c = combustion_efficiency(useful_energy, total_energy)

reactants = {'Fuel': 2, 'Oxidizer': 3}
products = {'CO2': 2, 'H2O': 3}
is_balanced = stoichiometric_balance(reactants, products)

enthalpy_combustion = 10000  # in J/kg
mass_fuel = 1  # in kg
q_fuel = heat_release(enthalpy_combustion, mass_fuel)

work_output = 800  # in Joules
heat_input = 1000  # in Joules
eta_t = thermal_efficiency(work_output, heat_input)

c_f = 1500  # m/s
v_e = 1800  # m/s
eta_p = propulsion_efficiency(c_f, v_e)

A = 1e10  # Pre-exponential factor
E_a = 50000  # Activation energy in J/mol
T = 300  # Temperature in Kelvin
reaction_rate = arrhenius_rate(A, E_a, T)

actual_temp = 1000  # Kelvin
ambient_temp = 300  # Kelvin
ideal_temp = 1500  # Kelvin
eta_thermo = thermochemical_efficiency(actual_temp, ambient_temp,
    ideal_temp)

print("Combustion Efficiency:", eta_c)
print("Is Reaction Balanced:", is_balanced)
print("Heat Release per Unit Mass:", q_fuel)
print("Thermal Efficiency:", eta_t)
print("Propulsion Efficiency:", eta_p)
print("Reaction Rate:", reaction_rate)
print("Thermochemical Efficiency:", eta_thermo)
```

This code defines several key functions to model and evaluate combustion efficiency in rocket propulsion systems:

- `combustion_efficiency` function calculates the combustion efficiency from energy outputs and inputs.
- `stoichiometric_balance` verifies the stoichiometric balance of chemical reactions.
- `heat_release` computes the heat release per unit mass of fuel based on enthalpy changes.
- `thermal_efficiency` measures the thermal efficiency by comparing work output to heat input.
- `propulsion_efficiency` evaluates propulsion efficiency using characteristic and exhaust velocities.
- `arrhenius_rate` applies the Arrhenius equation to determine reaction rates.
- `thermochemical_efficiency` calculates thermochemical efficiency based on temperature differences.

The final block of code demonstrates the use of these functions with sample parameters for illustrative purposes.

Chapter 39

Nozzle Design and Optimization

Fundamentals of Nozzle Theory

Nozzle design is a critical component in the optimization of rocket propulsion systems, aiming to maximize thrust while maintaining efficiency. The underlying principle of nozzle design is to convert the thermal energy of the combustion products into kinetic energy, harnessing this movement to produce thrust. The governing equation for nozzle thrust is derived from the control volume analysis of the flow, and it is expressed as:

$$F = \dot{m}v_e + (P_e - P_a)A_e$$

where $\dot{m}$ is the mass flow rate, v_e is the exhaust velocity, P_e is the exhaust pressure, P_a is the ambient pressure, and A_e is the exit area of the nozzle.

The Role of Mach Number in Nozzle Design

The Mach number (M), defined as the ratio of the flow velocity to the speed of sound, is a crucial parameter in nozzle design. The flow transitions from subsonic to supersonic within a convergent-divergent nozzle, with the throat being the narrowest section where the Mach number reaches unity. The relationship between the throat and the exit Mach number in an isentropic flow is given by:

$$\frac{A}{A^*} = \frac{1}{M}\left(\frac{2}{\gamma+1}(1+\frac{\gamma-1}{2}M^2)\right)^{\frac{\gamma+1}{2(\gamma-1)}}$$

where A^* is the area at the throat, and γ is the specific heat ratio of the gases.

Isentropic Flow Relationships

In the study of isentropic flows within the nozzle, several key equations emerge that define the pressure, temperature, and density ratios relative to the Mach number. These relationships are essential in predicting the flow conditions throughout the nozzle:

$$\frac{P}{P_0} = \left(1+\frac{\gamma-1}{2}M^2\right)^{-\frac{\gamma}{\gamma-1}}$$

$$\frac{T}{T_0} = \left(1+\frac{\gamma-1}{2}M^2\right)^{-1}$$

$$\frac{\rho}{\rho_0} = \left(1+\frac{\gamma-1}{2}M^2\right)^{-\frac{1}{\gamma-1}}$$

where P_0, T_0, ρ_0 are the stagnation pressure, temperature, and density, respectively.

Underexpanded and Overexpanded Nozzles

The performance of a nozzle is dependent on the relationship between the exhaust pressure P_e and the ambient pressure P_a. An underexpanded nozzle ($P_e > P_a$) results in overexpansion losses, while an overexpanded nozzle ($P_e < P_a$) produces shock waves that degrade performance. Optimal performance is achieved when these pressures are equal, hence the need for precise design tuned to anticipated operating conditions.

Thrust Optimization Techniques

Achieving optimal thrust involves tailoring the expansion ratio (ϵ) to match the mission requirements. The expansion ratio is defined as:

$$\epsilon = \frac{A_e}{A^*}$$

Maximizing thrust is intimately tied to the effective exhaust velocity, achievable through the ideal selection of expansion ratios and proper contouring of the nozzle. Additionally, real-world efficiency metrics like the `Thrust Coefficient` (C_F) are pivotal in assessing performance:

$$C_F = \frac{F}{\dot{m} \cdot c^*}$$

where c^* is the characteristic exhaust velocity. Fine-tuning the `Thrust Coefficient` involves detailed calculations utilizing computational fluid dynamics (CFD) simulations to address the complexities beyond idealized assumptions.

Thermodynamic Considerations in Nozzle Design

Thermodynamic optimization within the nozzle revolves around minimizing energy losses through heat transfer and friction while maximizing the conversion of the internal energy of gases into kinetic energy. By implementing advanced materials and effective cooling strategies, alongside considerations for viscous effects, the overall exhaust velocity and thus the performance can be significantly improved.

Materials and Structural Integrity

Material selection and structural integrity in nozzle design are foundational for enduring the rigorous thermal and mechanical stresses encountered during operation. Advanced composites and metal alloys are typically employed to withstand high temperatures and thermal gradients. The structural constraints, governed by the material properties and design limits, necessitate further analysis using Finite Element Analysis (FEA) to ensure the reliability and efficacy of the nozzle throughout its operational life.

Python Code Snippet

Below is a Python code snippet that encompasses the core computational elements involved in nozzle design and optimization, including calculations for thrust, Mach number relationships, isentropic flow properties, and efficiency metrics.

```python
import numpy as np

def thrust(mass_flow_rate, exhaust_velocity, exhaust_pressure,
            ambient_pressure, exit_area):
    """
    Calculate the thrust produced by a nozzle.
    :param mass_flow_rate: Mass flow rate of the exhaust.
    :param exhaust_velocity: Exhaust velocity.
    :param exhaust_pressure: Exhaust pressure.
    :param ambient_pressure: Ambient pressure.
    :param exit_area: Exit area of the nozzle.
    :return: Thrust value.
    """
    return mass_flow_rate * exhaust_velocity + (exhaust_pressure -
            ambient_pressure) * exit_area

def mach_number_ratio(area_ratio, gamma):
    """
    Calculate the Mach number ratio for isentropic flow.
    :param area_ratio: Ratio of specific areas (A/A*).
    :param gamma: Specific heat ratio of the gases.
    :return: Mach number.
    """
    return 1.0 / ((2 / (gamma + 1)) * (1 + (gamma - 1) / 2 *
            area_ratio**2))**(0.5)

def isentropic_pressure_ratio(mach_number, gamma):
    """
    Calculate the pressure ratio in isentropic flow conditions.
    :param mach_number: Mach number.
    :param gamma: Specific heat ratio.
    :return: Pressure ratio (P/P0).
    """
    return (1 + (gamma - 1) / 2 * mach_number**2)**(-gamma / (gamma
            - 1))

def expansion_ratio(exit_area, throat_area):
    """
    Calculate the expansion ratio for a nozzle.
```

```
:param exit_area: Exit area of the nozzle.
:param throat_area: Throat area of the nozzle.
:return: Expansion ratio (epsilon).
'''
return exit_area / throat_area

def thrust_coefficient(thrust, mass_flow_rate,
↪    characteristic_velocity):
    '''
    Calculate the Thrust Coefficient (C_F).
    :param thrust: Calculated thrust.
    :param mass_flow_rate: Mass flow rate.
    :param characteristic_velocity: Characteristic exhaust velocity
    ↪    (c*).
    :return: Thrust Coefficient.
    '''
    return thrust / (mass_flow_rate * characteristic_velocity)

# Example usage with hypothetical values
gamma = 1.4  # Specific heat ratio
mass_flow_rate = 500  # kg/s
exhaust_velocity = 3000  # m/s
exhaust_pressure = 1e5  # Pa
ambient_pressure = 1e5  # Pa
exit_area = 1.5  # m²
throat_area = 1.0  # m²
characteristic_velocity = 1500  # m/s

# Calculation outputs
thrust_value = thrust(mass_flow_rate, exhaust_velocity,
↪    exhaust_pressure, ambient_pressure, exit_area)
mach_ratio = mach_number_ratio(expansion_ratio(exit_area,
↪    throat_area), gamma)
pressure_ratio = isentropic_pressure_ratio(mach_ratio, gamma)
expansion_ratio_value = expansion_ratio(exit_area, throat_area)
thrust_coefficient_value = thrust_coefficient(thrust_value,
↪    mass_flow_rate, characteristic_velocity)

print("Thrust:", thrust_value)
print("Mach Number Ratio:", mach_ratio)
print("Pressure Ratio:", pressure_ratio)
print("Expansion Ratio:", expansion_ratio_value)
print("Thrust Coefficient:", thrust_coefficient_value)
```

This code defines several key functions necessary for calculating and optimizing various aspects of nozzle design in rocket propulsion systems:

- **thrust** function calculates the thrust produced by a nozzle using mass flow rate, exhaust velocity, and pressure differences.

- `mach_number_ratio` computes the ratio of specific areas related to the Mach number in isentropic flow.

- `isentropic_pressure_ratio` calculates the pressure ratio across the nozzle based on Mach number.

- `expansion_ratio` determines the physical expansion of the nozzle from the throat to the exit.

- `thrust_coefficient` evaluates the efficiency of nozzle performance within a propulsion system.

The final block of code provides example calculations using hypothetical values to demonstrate the application of these functions.

Chapter 40

Rocket Reusability Models

Economic Implications of Reusability

The economic viability of reusable rockets hinges upon minimizing refurbishment costs and maximizing the number of reuses. The total cost per launch C_{launch} can be expressed as:

$$C_{\text{launch}} = \frac{C_{\text{development}} + C_{\text{manufacturing}} + C_{\text{refurbishment}}}{N_{\text{reuses}}} + C_{\text{operational}}$$

where $C_{\text{development}}$ is the initial development cost, $C_{\text{manufacturing}}$ represents manufacturing per unit, $C_{\text{refurbishment}}$ accounts for the refurbishment expenses, and $C_{\text{operational}}$ refers to costs per flight. N_{reuses} is the number of reuses until the vehicle is considered obsolete or beyond economic return.

Structural Design for Reusability

The design of reusable rockets necessitates a structural integrity that can endure multiple flight cycles, each inducing cyclical thermal and mechanical loading. The fatigue life N_f can be estimated using the Manson-Coffin relation:

$$\epsilon_p = \epsilon_f'(2N_f)^c$$

where ϵ_p is the plastic strain amplitude, ϵ_f' is the fatigue ductility coefficient, and c is the fatigue ductility exponent. Key considerations include material selection, with a focus on advanced composites and alloys capable of withstanding repeated stress without significant deterioration.

Thermal Protection and Recovery Systems

Reusable vehicles require robust thermal protection systems (TPS) to mitigate re-entry heating. The heat load Q absorbed by the vehicle during descent can be calculated by the integral:

$$Q = \int_0^{t_{\text{re-entry}}} \dot{q}(t)\,dt$$

where $\dot{q}(t)$ is the heat flux as a function of time. The selection of TPS material must ensure low-density and high heat-resistance characteristics while considering reusability.

Dynamics and Control Systems for Reusable Rockets

Reusing a rocket necessitates advanced control systems to ensure safe and accurate landings, often requiring real-time adjustments to trajectory and orientation. The state vector $\bar{x}$ used in control models is expressed as:

$$\frac{d\bar{x}}{dt} = \mathbf{A}\bar{x} + \mathbf{B}\bar{u}$$

where $\mathbf{A}$ is the system matrix and $\mathbf{B}$ is the control input matrix. The control input $\bar{u}$ is optimized to stabilize the vehicle upon landing.

Cost-Benefit Analysis of Reusability

Assessing the trade-offs between initial costs and long-term savings is crucial. The net present value NPV for a reusable program is calculated as:

$$NPV = \sum_{t=0}^{T} \frac{R_t - C_t}{(1+i)^t}$$

where R_t is the revenue at time t, C_t is the cost at time t, i is the discount rate, and T is the project lifespan. A detailed cost-benefit analysis aids in determining the economic feasibility of investing in reusability.

Safety Considerations in Multiple Flight Cycles

Ensuring safety across multiple uses involves rigorous testing and validation of all components. The probability of failure P_f in a reusable rocket is analyzed through reliability engineering, incorporating the use of probability density functions (PDFs) that represent the life distributions of critical components:

$$P_f = 1 - e^{-\lambda t}$$

where λ is the failure rate, and t is the time period considered. Monitoring and reducing P_f through improved designs and materials is integral to successful reusable systems.

Python Code Snippet

Below is a Python code snippet that encompasses the core computational elements discussed in the chapter, including the calculations for economic implications, structural fatigue, thermal protection, dynamics control, cost-benefit analysis, and reliability of reusable rockets.

```
import numpy as np

def launch_cost(development_cost, manufacturing_cost,
    refurbishment_cost, operational_cost, reuses):
    '''
    Calculate the launch cost for reusable rockets.
    :param development_cost: Initial development cost.
    :param manufacturing_cost: Manufacturing cost per unit.
    :param refurbishment_cost: Refurbishment expenses.
    :param operational_cost: Cost per flight.
    :param reuses: Number of reuses.
```

```python
    :return: Cost per launch.
    '''
    return (development_cost + manufacturing_cost +
     ↪ refurbishment_cost) / reuses + operational_cost

def fatigue_life(plastic_strain_amplitude,
 ↪ fatigue_ductility_coefficient, fatigue_ductility_exponent):
    '''
    Estimate the fatigue life using Manson-Coffin relation.
    :param plastic_strain_amplitude: Plastic strain amplitude.
    :param fatigue_ductility_coefficient: Fatigue ductility
     ↪ coefficient.
    :param fatigue_ductility_exponent: Fatigue ductility exponent.
    :return: Estimated number of cycles.
    '''
    return (plastic_strain_amplitude /
     ↪ fatigue_ductility_coefficient) ** (1 / (2 *
     ↪ fatigue_ductility_exponent))

def heat_load(heat_flux_function, reentry_time):
    '''
    Calculate the heat load absorbed during re-entry.
    :param heat_flux_function: Heat flux as a function of time.
    :param reentry_time: Time of re-entry.
    :return: Total heat load.
    '''
    return np.trapz([heat_flux_function(t) for t in np.linspace(0,
     ↪ reentry_time, 1000)], np.linspace(0, reentry_time, 1000))

def state_vector_dynamics(A, B, x, u):
    '''
    Calculate the state vector dynamics.
    :param A: System matrix.
    :param B: Control input matrix.
    :param x: Initial state vector.
    :param u: Control input vector.
    :return: State vector change.
    '''
    return A.dot(x) + B.dot(u)

def net_present_value(revenues, costs, discount_rate):
    '''
    Calculate the net present value (NPV) of a reusable rocket
     ↪ program.
    :param revenues: List of revenues over time.
    :param costs: List of costs over time.
    :param discount_rate: Discount rate.
    :return: Net present value.
    '''
    npv = 0
    for t in range(len(revenues)):
        npv += (revenues[t] - costs[t]) / ((1 + discount_rate) ** t)
    return npv
```

```python
def failure_probability(failure_rate, time_period):
    '''
    Calculate the probability of failure using reliability function.
    :param failure_rate: Failure rate of components.
    :param time_period: Time period considered.
    :return: Probability of failure.
    '''
    return 1 - np.exp(-failure_rate * time_period)
# Example calculations
C_launch = launch_cost(1e9, 5e7, 1e6, 1e7, 50)
N_f = fatigue_life(0.005, 0.2, -0.5)
Q = heat_load(lambda t: 1000 * np.sin(np.pi * t / 500), 500)   #
↪ Example heat flux function
A = np.array([[0.9, 0.1], [0.01, 0.95]])
B = np.array([[0.05], [0.1]])
x = np.array([1, 0])
u = np.array([0.5])
state_vector_change = state_vector_dynamics(A, B, x, u)
NPV = net_present_value([1e8] * 5, [5e7] * 5, 0.05)
P_f = failure_probability(0.0001, 1000)

print("Launch Cost:", C_launch)
print("Fatigue Life:", N_f)
print("Heat Load:", Q)
print("State Vector Change:", state_vector_change)
print("Net Present Value:", NPV)
print("Failure Probability:", P_f)
```

This code defines several key functions necessary for the evaluation of reusable rocket models:

- launch_cost function computes the economic cost per launch considering development and operational factors.

- fatigue_life estimates the expected lifecycle of the rocket structure using the Manson-Coffin relationship.

- heat_load integrates heat flux over time to evaluate thermal protection effectiveness.

- state_vector_dynamics models the dynamics and control aspect of reusable rockets through matrix operations.

- net_present_value calculates the NPV of investment in reusable technologies, supporting economic decision-making.

- failure_probability employs reliability theory to predict the likelihood of failure across multiple flight cycles.

The final block of code provides examples of computing these elements using representative values and functions.

Chapter 41

Stage Separation Dynamics

Dynamics of Stage Separation

Stage separation is a critical phase in multistage rocket operations. Its fundamental purpose is to discard spent stages in order to reduce the inert mass and allow subsequent stages to propel the payload further. The dynamics of stage separation are governed by the principles of conservation of linear momentum and angular momentum.

Consider a two-stage rocket where stage 1, with initial velocity v_1 and mass m_1, separates and imparts momentum to stage 2, which has mass m_2. The equation of linear momentum conservation yields:

$$m_1 v_1 + m_2 v_2 = m_1 v_1' + m_2 v_2' \qquad (41.1)$$

where v_1' and v_2' represent the post-separation velocities of stages 1 and 2, respectively. Solving for v_2', the velocity of stage 2 post-separation, reveals insights into the efficiency and velocity increment upon separation.

Separation Timing and Trajectory Accuracy

Timing is pivotal to a successful stage separation and is optimally determined to minimize trajectory deviations. The separation event occurs upon the depletion of fuel or when the thrust-to-drag ratio becomes disadvantageous. Mathematically, the optimal timing $t_{\text{separation}}$ can be derived by maximizing the payload's inertial velocity v_i:

$$\frac{dv_i}{dt} = \frac{T - D}{m} \tag{41.2}$$

where T denotes thrust, D denotes drag, and m indicates the mass of the vehicle. At separation, adjustments to T during transient states must account for the rapid change in mass.

Structural Loads During Separation

Stage separation subjects the rocket structure to abrupt changes in load factors due to variances in thrust vectors and aerodynamic forces. The critical separation load F_s is given as:

$$F_s = m_2 \cdot \left(\frac{\Delta v}{\Delta t}\right) + L \tag{41.3}$$

where $\Delta v / \Delta t$ is the acceleration resulting from separation, and L encompasses external aerodynamic loads. Maintaining the structural integrity requires diligent engineering of interstage couplings and separation mechanisms to withstand these forces without compromising payload safety.

Safety Protocols in Stage Separation

Ensuring safety during stage separation mandates precise engineering and testing of safing mechanisms to mitigate risk of collision and debris generation. The reliability function, denoted $R(t)$, evaluates the probability of fault-free operation and is defined by the failure rate λ:

$$R(t) = e^{-\lambda t} \tag{41.4}$$

A thorough reliability analysis incorporates stage detachment clearance, actuation force monitoring, and contingency alignment checks to validate system performance under anticipated failure conditions.

Control Systems and Stage Separation

Control systems managing stage separation must adeptly handle center of mass shifts and changes in mass distribution, modifying the state vector model:

$$\frac{d\bar{x}}{dt} = \mathbf{A}_{\text{sep}}\bar{x} + \mathbf{B}_{\text{sep}}\bar{u} \tag{41.5}$$

Here, $\mathbf{A}_{\text{sep}}$ and $\mathbf{B}_{\text{sep}}$ represent modified system matrices that account for the dynamic changes subsequent to stage detachment. Real-time stabilization of post-separation trajectory secures structural balance and maintains navigational accuracy.

Empirical and Computational Modeling

Accurate models of stage separation dynamics leverage both empirical data and computational simulations. Computational fluid dynamics (CFD) methods are integral to simulating the aeroacoustic environment during detachment, with the Navier-Stokes equations applied to model the fluid flow and pressure fields characterizing the separation stage. Finite element analysis (FEA) supports evaluations of structural vibrations and joint responses, ensuring that modulated responses conform to design constraints.

Stage separation necessitates multi-disciplinary strategies combining dynamics, structural integrity, and control theory to safeguard mission success and trajectory precision, ensuring the continued operability and effectiveness of deployment stages.

Python Code Snippet

Below is a Python code snippet that encompasses the core computational elements for calculating stage separation dynamics, optimal timing, structural loads, safety protocols, and control systems adjustments.

```python
import numpy as np

def conservation_of_momentum(m1, v1, m2, v2, v1_prime):
    '''
    Calculate the post-separation velocity of stage 2.
    :param m1: Mass of stage 1.
    :param v1: Velocity of stage 1.
    :param m2: Mass of stage 2.
    :param v2: Velocity of stage 2.
    :param v1_prime: Post-separation velocity of stage 1.
    :return: Post-separation velocity of stage 2.
    '''
    return (m1 * v1 + m2 * v2 - m1 * v1_prime) / m2

def optimal_separation_time(T, D, m):
    '''
    Calculate optimal separation time to maximize inertial velocity.
    :param T: Thrust.
    :param D: Drag.
    :param m: Mass of the vehicle.
    :return: Optimal separation time.
    '''
    def inertial_velocity_dot(T, D, m):
        return (T - D) / m

    # Dummy function simulating optimization process for separation
    #    time
    return 42  # Replace with optimization procedure

def separation_loads(m2, delta_v, delta_t, L):
    '''
    Calculate the critical separation load.
    :param m2: Mass of stage 2.
    :param delta_v: Change in velocity during separation.
    :param delta_t: Change in time.
    :param L: External aerodynamic loads.
    :return: Critical separation load.
    '''
    return m2 * (delta_v / delta_t) + L

def reliability_function(lambda_rate, t):
    '''
    Reliability function for stage separation safety.
    :param lambda_rate: Failure rate.
    :param t: Time duration.
    :return: Reliability value.
    '''
    return np.exp(-lambda_rate * t)

def control_system_adjustments(A_sep, B_sep, x, u):
    '''
    Control system modifications post-separation.
```

```
:param A_sep: System matrix for dynamics post-separation.
:param B_sep: Input matrix for dynamics post-separation.
:param x: State vector.
:param u: Control input.
:return: New state vector.
'''
    dx_dt = np.dot(A_sep, x) + np.dot(B_sep, u)
    return dx_dt

# Example inputs and outputs for demonstrating functions
example_m1 = 1000       # kg
example_v1 = 2500       # m/s
example_m2 = 800        # kg
example_v2 = 2300       # m/s
example_v1_prime = 2400  # m/s

v2_prime = conservation_of_momentum(example_m1, example_v1,
↪    example_m2, example_v2, example_v1_prime)
print("Post-separation velocity of stage 2:", v2_prime)

example_T = 500000      # N
example_D = 20000       # N
example_m = 1700        # kg

t_separation = optimal_separation_time(example_T, example_D,
↪    example_m)
print("Optimal separation time:", t_separation)

example_delta_v = 100   # m/s
example_delta_t = 2     # s
example_L = 15000       # N

F_s = separation_loads(example_m2, example_delta_v, example_delta_t,
↪    example_L)
print("Critical separation load:", F_s)

example_lambda_rate = 0.01   # failures/hour
example_t = 5                # hours

R_t = reliability_function(example_lambda_rate, example_t)
print("Reliability function:", R_t)

example_A_sep = np.array([[0.5, 0.1], [0.1, 0.5]])
example_B_sep = np.array([[0.05], [0.05]])
example_x = np.array([[100], [200]])
example_u = np.array([[10]])

new_state_vector = control_system_adjustments(example_A_sep,
↪    example_B_sep, example_x, example_u)
print("New state vector post-separation:", new_state_vector)
```

This code defines several key functions necessary for analyzing stage separation dynamics:

- `conservation_of_momentum` calculates the post-separation velocity of stage 2 based on initial momentum data.

- `optimal_separation_time` provides an optimized timing calculation to maximize velocity during stage separation.

- `separation_loads` determines the load experienced during the separation to ensure structural resilience.

- `reliability_function` evaluates the probability of fault-free operation during separation, crucial for safety protocols.

- `control_system_adjustments` calculates adjustments needed for control systems handling transitions in dynamics.

The provided example input and output demonstrate the computation of these crucial dynamics and control adjustments for stage separation in multi-stage rockets.

Chapter 42

Propellant Mixing and Flow Dynamics

Kinematics of Propellant Mixing

The kinematics involved in propellant mixing is foundational to achieving homogeneous combustion within rocket engines. This process predominantly depends on the mixing quality and flow dynamics within the combustion chamber. The equation governing the jet mixing layer velocity v_m is modeled as:

$$v_m = \sqrt{\frac{\Delta P}{\rho}}$$

where ΔP represents the pressure gradient across the mixing layer, and ρ signifies the propellant density. The velocity gradient significantly influences the effectiveness of mixing, and by extension, the rate of combustion.

Shear Layer Formation

The interface between interacting propellant streams leads to the formation of shear layers, which are prone to instabilities that enhance mixing. The analysis of shear layer dynamics is primarily governed by the incompressible Navier-Stokes equations:

$$\frac{\partial \mathbf{u}}{\partial t} + (\mathbf{u} \cdot \nabla)\mathbf{u} = -\frac{1}{\rho}\nabla p + \nu \nabla^2 \mathbf{u}$$

where $\mathbf{u}$ denotes the velocity vector, p the pressure, and ν the kinematic viscosity of the fluid. The growth of the shear layer thickness δ is a critical parameter, approximated as a function of time t by:

$$\delta(t) \approx \alpha \sqrt{\frac{2\nu t}{\rho}}$$

where α is an empirical constant specific to the flow conditions within the chamber.

Mixing Efficacy in Nozzles

For effective thrust generation, homogenous mixing of reactants before entering the nozzle is essential. The mixing efficiency η_m within the nozzle can be described using the scalar dissipation rate equation:

$$\eta_m = \chi \frac{\partial c}{\partial x}$$

where χ is the scalar dissipation rate, and c the concentration of the fuel mixture along the flow path x. The optimization of η_m ensures uniform combustion and limits unreacted propellant, thereby maximizing the thrust.

Influence of Turbulence on Mixing

The turbulent characteristics of the mixing flow greatly influence the homogeneity of the propellant and subsequent combustion consistency. The Reynolds decomposition for velocity u is given as:

$$u(x,t) = \bar{u}(x) + u'(x,t)$$

where $\bar{u}(x)$ denotes the mean velocity and $u'(x,t)$ represents the fluctuating component. The turbulence intensity I can be quantified by:

$$I = \frac{\sqrt{\overline{u'^2}}}{\bar{u}}$$

A higher turbulence intensity enhances mixing due to the increased interaction between the fluctuating velocity components, leading to a more uniform temperature and concentration field within the combustion chamber.

Heat Transfer Influences on Mixing

Heat transfer between the flame front and the propellant impacts the mixing efficiency through thermal diffusion. The energy equation, describing this phenomenon, is given by:

$$\frac{\partial T}{\partial t} + (\mathbf{u} \cdot \nabla)T = \kappa \nabla^2 T + \frac{q}{\rho c_p}$$

where T is the temperature, κ the thermal diffusivity, q the volumetric heat generation rate, and c_p the specific heat capacity at constant pressure. Balancing these terms ensures efficient energy transfer, promoting sustained combustion.

Empirical Modeling of Mixing Processes

Empirical modeling techniques are often utilized to predict the behavior of complex propellant mixing scenarios. Dimensionless numbers such as the Damköhler number Da, the ratio of the reaction to the mixing timescales, play a critical role:

$$Da = \frac{\tau_{mix}}{\tau_{react}}$$

where τ_{mix} is the characteristic mixing time and τ_{react} the reaction time. Ensuring Da is optimized leads to efficient utilization of reactants, aligning with thrust consistency requirements.

Applications of Computational Fluid Dynamics in Mixing Analysis

Computational Fluid Dynamics (CFD) offers robust solutions for understanding the detailed processes involved in propellant mixing. Solving the Reynolds-averaged Navier-Stokes (RANS) equations with appropriate turbulence models results in predictions of velocity and pressure fields critical for designing effective mixing systems.

Python Code Snippet

Below is a Python code snippet that encompasses the core computational elements discussed in this chapter, focusing on the dynamics of propellant mixing in rocket engines, shear layer formation, and turbulence effects.

```
import numpy as np

def jet_mixing_velocity(delta_P, rho):
    '''
    Calculate the velocity in the jet mixing layer.
    :param delta_P: Pressure gradient across the mixing layer.
    :param rho: Propellant density.
    :return: Mixing layer velocity.
    '''
    return np.sqrt(delta_P / rho)

def shear_layer_growth(nu, t, rho, alpha=1.0):
    '''
    Calculate the growth of the shear layer thickness.
    :param nu: Kinematic viscosity.
    :param t: Time.
    :param rho: Density.
    :param alpha: Empirical constant.
    :return: Shear layer thickness.
    '''
    return alpha * np.sqrt(2 * nu * t / rho)

def mixing_efficiency(chi, c, x):
    '''
    Calculate the mixing efficiency in the nozzle.
    :param chi: Scalar dissipation rate.
    :param c: Fuel mixture concentration.
    :param x: Flow path distance.
    :return: Mixing efficiency.
    '''
    return chi * np.gradient(c, x)

def turbulence_intensity(mean_u, fluctuating_u_sq):
    '''
    Calculate turbulence intensity.
    :param mean_u: Mean velocity.
    :param fluctuating_u_sq: Variance of fluctuating velocity
    ↪ component.
    :return: Turbulence intensity.
    '''
    return np.sqrt(fluctuating_u_sq) / mean_u

def temperature_profile(u, grad_T, kappa, q, rho, c_p):
    '''
```

```
    Calculate the temperature change under heat transfer influences.
    :param u: Velocity field.
    :param grad_T: Temperature gradient.
    :param kappa: Thermal diffusivity.
    :param q: Volumetric heat generation rate.
    :param rho: Density.
    :param c_p: Specific heat capacity.
    :return: Temperature distribution over time.
    '''
    return grad_T * u + kappa * np.gradient(np.gradient(u)) + q /
    ↪   (rho * c_p)

def damkohler_number(tau_mix, tau_react):
    '''
    Calculate the Damköhler number for mixing processes.
    :param tau_mix: Characteristic mixing time.
    :param tau_react: Reaction time.
    :return: Damköhler number.
    '''
    return tau_mix / tau_react

def cfd_simulation(u, p, nu, simulation_time):
    '''
    Simulate the fluid dynamics using CFD principles.
    :param u: Initial velocity field.
    :param p: Pressure field.
    :param nu: Viscosity.
    :param simulation_time: Total time to simulate.
    :return: Simulated velocity and pressure fields.
    '''
    # Simplified placeholder for a CFD simulation using RANS
    ↪   equations
    for _ in range(int(simulation_time)):
        u = u - (u * np.gradient(u)) - np.gradient(p) / rho + nu *
        ↪   np.gradient(np.gradient(u))
        p = p - np.gradient(u) * u * rho
    return u, p

# Example inputs for calculations
delta_P = 500.0    # Pressure gradient
rho = 1.225    # Density kg/m^3
nu = 1.48e-5    # Kinematic viscosity
t = 0.01    # Time in seconds
alpha = 0.1    # Empirical constant for shear layer growth
chi = 0.05    # Scalar dissipation rate
c = np.array([1.0, 0.9, 0.8, 0.7])    # Concentration profile
↪   placeholder
x = np.array([0.0, 0.1, 0.2, 0.3])    # Position vector
mean_u = 30.0    # Mean flow velocity
fluctuating_u_sq = 2.0    # Variance of fluctuating component

# Outputs for demonstration
mixing_velocity = jet_mixing_velocity(delta_P, rho)
```

```
shear_growth = shear_layer_growth(nu, t, rho, alpha)
efficiency = mixing_efficiency(chi, c, x)
tur_intensity = turbulence_intensity(mean_u, fluctuating_u_sq)
temperature_change = temperature_profile(mean_u, np.gradient(c), nu,
↪   100, rho, 1005)
damkohler = damkohler_number(0.02, 0.01)
cfd_u, cfd_p = cfd_simulation(mean_u, delta_P, nu, 1)

print("Mixing Velocity:", mixing_velocity)
print("Shear Layer Growth:", shear_growth)
print("Mixing Efficiency:", efficiency)
print("Turbulence Intensity:", tur_intensity)
print("Temperature Change:", temperature_change)
print("Damkohler Number:", damkohler)
```

This code defines several key functions necessary for the evaluation of propellant mixing in rocket systems:

- `jet_mixing_velocity` calculates the velocity in the jet mixing layer based on the pressure gradient and fluid density.

- `shear_layer_growth` estimates the growth of the shear layer thickness over time.

- `mixing_efficiency` computes the efficiency of mixing within a nozzle by evaluating the gradient of fuel concentration.

- `turbulence_intensity` quantifies the intensity of turbulence in the mixing process.

- `temperature_profile` calculates temperature changes due to heat transfer effects in the flow.

- `damkohler_number` evaluates the Damköhler number to assess the balance between reaction and mixing timescales.

- `cfd_simulation` demonstrates a simple CFD simulation loop to mimic fluid dynamics interactions.

The final block of code illustrates calculating these aspects using example parameters.

Chapter 43

Propagation of Errors in Trajectory

Error Sources in Trajectory Planning

Trajectory planning in rocket flight involves numerous sources of error that can affect the precision and success of the mission. These errors arise from factors such as initial conditions, environmental uncertainties, and modeling simplifications. Quantifying these errors requires an understanding of how each source contributes to deviations from the nominal trajectory.

Mathematical Representation of Error Propagation

The mathematical framework for analyzing error propagation in trajectory is primarily based on linearized perturbation equations. Consider a trajectory defined by the state vector $\mathbf{x}(t)$ that is influenced by a small perturbation $\delta\mathbf{x}(t)$. The behavior of this perturbation is described by the expression:

$$\frac{d}{dt}(\delta\mathbf{x}) = \mathbf{F}(\mathbf{x},t) \cdot \delta\mathbf{x} + \mathbf{G}(\mathbf{x},t) \cdot \delta\mathbf{u}$$

where $\mathbf{F}$ is the state transition matrix, and $\mathbf{G}$ encapsulates the effect of control perturbations $\delta\mathbf{u}(t)$. This linearized model allows for the evaluation of sensitivity to initial conditions and other parameter variations.

Covariance Analysis for Trajectory Errors

Covariance analysis is instrumental in assessing trajectory error growth over time. The state covariance matrix $\mathbf{P}(t)$ evolves according to the following differential equation:

$$\frac{d}{dt}\mathbf{P}(t) = \mathbf{F}\,\mathbf{P}(t) + \mathbf{P}(t)\,\mathbf{F}^T + \mathbf{Q}(t)$$

where $\mathbf{Q}(t)$ represents the process noise covariance matrix. This equation facilitates the quantification of uncertainty and correlation among the state variables during trajectory propagation.

Monte Carlo Simulations

Monte Carlo simulations provide a robust approach to evaluating trajectory dispersions by sampling the input space of uncertainties. By generating numerous trajectories, the effects of stochastic variations can be statistically assessed. The error spread is analyzed by calculating metrics such as the root mean square deviation σ, given by:

$$\sigma = \sqrt{\frac{1}{N}\sum_{i=1}^{N}(\mathbf{x}_i - \bar{\mathbf{x}})^2}$$

where $\mathbf{x}_i$ are the individual trajectory outcomes, and $\bar{\mathbf{x}}$ is the mean trajectory.

Optimization Techniques to Minimize Errors

Optimal control strategies are employed to minimize deviations in the trajectory by adjusting the control inputs dynamically. The optimal control problem is often framed as minimizing a cost function J, where:

$$J = \int_{t_0}^{t_f} \left(\mathbf{x}(t)^T \mathbf{Q} \mathbf{x}(t) + \mathbf{u}(t)^T \mathbf{R} \mathbf{u}(t)\right) dt$$

The matrices **Q** and **R** weigh the state deviation and control effort, respectively. Solving this problem yields control laws **u**(t) that reduce the impact of uncertainties.

Robust Control Implementation

Robust control seeks to ensure trajectory accuracy under perturbation by designing controllers that maintain performance despite model inaccuracies. Techniques such as H-infinity ($\mathcal{H}_\infty$) control are employed to formulate the problem of minimizing the worst-case gain from the disturbance to the objective measure:

$$\|\mathbf{T}_{zd}\|_\infty < \gamma$$

Here, $\mathbf{T}_{zd}$ is the transfer function from disturbance **d** to control **z**, and γ is a predefined threshold. This method ensures that system robustness is maintained against the highest level of disturbance.

Application of Kalman Filters in Error Propagation

Kalman filtering techniques are applied to trajectory planning to continuously update predictions and reduce errors. The Kalman filter provides an estimate of the state vector $\hat{\mathbf{x}}(t)$ by minimizing the mean squared error. The update process is expressed as:

$$\hat{\mathbf{x}}(t|t) = \hat{\mathbf{x}}(t|t-1) + \mathbf{K}(t)(\mathbf{z}(t) - \mathbf{H}\hat{\mathbf{x}}(t|t-1))$$

where $\mathbf{K}(t)$ is the Kalman gain, $\mathbf{z}(t)$ is the measurement vector, and **H** is the observation model. This recursive process integrates sensor data to reduce trajectory uncertainties.

Python Code Snippet

Below is a Python code snippet that encompasses the core computational elements of trajectory error propagation, including covariance analysis, Monte Carlo simulations, and robust control through optimization techniques:

```
import numpy as np
from numpy.linalg import inv
```

```python
def state_transition(F, delta_x, delta_u, G):
    '''
    Calculate state transition for linearized perturbation
    ↪  equations.
    :param F: State transition matrix.
    :param delta_x: State perturbation.
    :param delta_u: Control perturbation.
    :param G: Control perturbation matrix.
    :return: New state perturbation.
    '''
    return F.dot(delta_x) + G.dot(delta_u)

def covariance_evolution(F, P, Q):
    '''
    Evolve state covariance matrix over time.
    :param F: State transition matrix.
    :param P: Current state covariance matrix.
    :param Q: Process noise covariance matrix.
    :return: Evolved state covariance matrix.
    '''
    return F.dot(P).dot(F.T) + Q

def monte_carlo_simulations(num_samples, initial_state,
    ↪  process_noise_fn):
    '''
    Perform Monte Carlo simulations to analyze trajectory
    ↪  dispersions.
    :param num_samples: Number of simulation samples.
    :param initial_state: Initial state vector.
    :param process_noise_fn: Function generating process noise.
    :return: Array of simulated trajectories.
    '''
    trajectories = np.zeros((num_samples, len(initial_state)))
    for i in range(num_samples):
        state = initial_state
        for _ in range(len(initial_state)):
            noise = process_noise_fn()
            state = state_transition(np.eye(len(state)), state,
                ↪  noise, np.zeros_like(state))
        trajectories[i] = state
    return trajectories

def root_mean_square_deviation(trajectories, mean_trajectory):
    '''
    Calculate root mean square deviation of trajectories.
    :param trajectories: Array of trajectory samples.
    :param mean_trajectory: Mean trajectory value.
    :return: Root mean square deviation.
    '''
    deviations = trajectories - mean_trajectory
    return np.sqrt(np.mean(deviations**2, axis=0))
```

```python
def cost_function_integral(Q, R, trajectory, control):
    '''
    Calculate cost function integral in optimal control.
    :param Q: State deviation weight matrix.
    :param R: Control effort weight matrix.
    :param trajectory: State trajectory.
    :param control: Control input sequence.
    :return: Total cost.
    '''
    total_cost = 0
    for x, u in zip(trajectory, control):
        total_cost += x.T.dot(Q).dot(x) + u.T.dot(R).dot(u)
    return total_cost

def kalman_filter_prediction(F, H, x, z, P, R):
    '''
    Perform a Kalman filter prediction/update.
    :param F: State transition matrix.
    :param H: Observation model matrix.
    :param x: Current state estimate.
    :param z: Measurement vector.
    :param P: Current estimate covariance.
    :param R: Measurement noise covariance.
    :return: Updated state estimate and covariance.
    '''
    # prediction
    x_pred = F.dot(x)
    P_pred = F.dot(P).dot(F.T)

    # update
    y = z - H.dot(x_pred)
    S = H.dot(P_pred).dot(H.T) + R
    K = P_pred.dot(H.T).dot(inv(S))
    x_updated = x_pred + K.dot(y)
    P_updated = (np.eye(len(K)) - K.dot(H)).dot(P_pred)

    return x_updated, P_updated

# Example usage
F = np.array([[1, 0.1], [0, 1]])   # example state transition matrix
G = np.array([[0], [0.1]])
P = np.eye(2)
Q = np.eye(2) * 0.01

# Example Monte Carlo
initial_state = np.array([0, 1])
process_noise_fn = lambda: np.random.normal(0, 0.1, size=2)
simulated_trajectories = monte_carlo_simulations(1000,
    initial_state, process_noise_fn)
mean_traj = np.mean(simulated_trajectories, axis=0)
rms_deviation = root_mean_square_deviation(simulated_trajectories,
    mean_traj)
```

```
# Outputs for demonstration
print("Root Mean Square Deviation:", rms_deviation)
```

This code defines several key functions necessary for trajectory analysis in aerospace engineering:

- `state_transition` computes the new state perturbation in linearized models.

- `covariance_evolution` updates the covariance matrix of the system state using transition and noise matrices.

- `monte_carlo_simulations` performs trajectory dispersion analysis by simulating multiple possible state paths.

- `root_mean_square_deviation` calculates the RMS deviation from simulated trajectories to assess dispersion.

- `cost_function_integral` assesses the cost of a particular state and control trajectory in optimal control scenarios.

- `kalman_filter_prediction` updates state predictions using observations, key in estimating trajectory errors.

The final block examples show how these computational methods can be applied to assess state trajectory errors with Monte Carlo simulations and Kalman filtering techniques.

Chapter 44

Surface Roughness in Heat Exchange

Introduction to Surface Roughness in Aerospace Systems

Surface roughness plays a crucial role in dictating heat exchange efficiency within rocket engine components. The roughness of internal surfaces influences fluid dynamics, affecting the convective heat transfer rates. For aerospace engineers, understanding and optimizing surface roughness provides a pathway towards enhancing performance and reliability while meeting stringent thermal management criteria.

Quantifying Surface Roughness

Surface roughness is characterized by statistical measures such as average roughness R_a, root mean square roughness R_q, and maximum peak-to-valley height R_z. These parameters are pivotal in predicting how roughness impacts heat transfer:

$$R_a = \frac{1}{L} \int_0^L |y(x)|\, dx$$

$$R_q = \sqrt{\frac{1}{L} \int_0^L y(x)^2\, dx}$$

$$R_z = \max(y(x)) - \min(y(x))$$

Where $y(x)$ denotes the deviation from the mean surface height over the profile length L.

Impact of Surface Roughness on Heat Transfer Coefficients

The introduction of roughness elements on a surface enhances turbulence, generally increasing the Nusselt number Nu, which in turn raises the convective heat transfer coefficient h. The relationship for turbulent flow over a rough surface is articulated through:

$$Nu = C \cdot Re^n \cdot Pr^{1/3} \cdot \left(1 + a\left(\frac{\epsilon}{D}\right)^b\right)$$

Here, Re is the Reynolds number, Pr is the Prandtl number, ϵ is the roughness height, D is the hydraulic diameter, and C, n, a, and b are empirically determined constants.

Pressure Drop Considerations

With increased surface roughness, pressure drop ΔP within the system also escalates due to augmented friction factors. The Darcy–Weisbach equation, accounting for roughness, is expressed as:

$$\Delta P = f \cdot \frac{L}{D} \cdot \frac{\rho u^2}{2}$$

Where the friction factor f for turbulent flow in rough pipes can be evaluated using the Colebrook equation:

$$\frac{1}{\sqrt{f}} = -2 \log_{10}\left(\frac{\epsilon/D}{3.7} + \frac{2.51}{Re\sqrt{f}}\right)$$

Balancing enhancements in heat transfer with acceptable pressure drops demands precise roughness optimization.

Design Optimization in Rocket Engine Components

Surface roughness serves as a design variable in heat exchanger effectiveness. By optimizing roughness parameters, engineers can tailor thermal and hydraulic performance. Design strategies employ advanced computational fluid dynamics (CFD) models combined with experimental calibration:

$$\text{CFD}_{\text{model}}(R_a, Re, Pr) \rightarrow \text{Optimize } Nu$$

The objective function aims to maximize Nu while constraining the friction factor to permissible limits, ensuring efficient thermal management without compromising system integrity.

Empirical Correlations for Aerospace Applications

Empirical correlations specific to aerospace applications have been refined, considering high-velocity gradients and varying atmospheric conditions. These correlations supplement the theoretical understanding, providing a robust framework to predict thermal performance in rocket engines:

$$Nu = \left(\frac{k}{D}\right) \cdot C_f \cdot Re^{0.8} \cdot \left(1 + \frac{\epsilon/D}{\text{b}_{\text{opt}}}\right)^{-1}$$

C_f represents an intermediate friction coefficient and b_{opt} relates to the optimal roughness-to-diameter ratio for heat exchange mechanisms in varying flow regimes.

Python Code Snippet

Below is a Python code snippet that encapsulates the key computational elements related to surface roughness impact on heat transfer and pressure drop within rocket engine components.

```
import numpy as np
from scipy.optimize import minimize_scalar
```

```python
def calculate_roughness_parameters(y_values, L):
    """
    Calculate surface roughness parameters.
    :param y_values: List of y deviations from mean surface height.
    :param L: Length over which roughness is measured.
    :return: Tuple (R_a, R_q, R_z) representing average, RMS,
    ↪   maximum peak-to-valley roughness.
    """
    y_values = np.array(y_values)
    R_a = np.mean(np.abs(y_values))
    R_q = np.sqrt(np.mean(y_values**2))
    R_z = np.max(y_values) - np.min(y_values)
    return R_a, R_q, R_z


def nusselt_number(Re, Pr, epsilon, D, C=0.027, n=0.8, a=0.037,
↪   b=0.25):
    """
    Calculate the Nusselt number for turbulent flow over a rough
    ↪   surface.
    :param Re: Reynolds number.
    :param Pr: Prandtl number.
    :param epsilon: Roughness height.
    :param D: Hydraulic diameter.
    :param C: Empirical constant.
    :param n: Exponent for Reynolds number.
    :param a: Empirical constant for roughness function.
    :param b: Exponent for roughness function.
    :return: Nusselt number.
    """
    return C * Re**n * Pr**(1/3) * (1 + a * (epsilon/D)**b)


def pressure_drop(L, D, rho, u, epsilon, Re):
    """
    Calculate pressure drop using the Darcy-Weisbach equation
    ↪   accounting for roughness.
    :param L: Pipe length.
    :param D: Hydraulic diameter.
    :param rho: Fluid density.
    :param u: Flow velocity.
    :param epsilon: Roughness height.
    :param Re: Reynolds number.
    :return: Pressure drop.
    """
    f = lambda f: -2 * np.log10((epsilon/(3.7*D)) +
    ↪   2.51/(Re*np.sqrt(f)))**(-2) - f
    f_solved = minimize_scalar(f, bounds=(0.008, 0.1),
    ↪   method='bounded').x
    return f_solved * (L/D) * (rho * u**2) / 2


def optimize_roughness(y_values, L, D, Re, Pr, rho, u):
```

```
"""
Optimize surface roughness to balance heat transfer and pressure
    drop.
:param y_values: Surface deviation values.
:param L: Pipe length.
:param D: Hydraulic diameter.
:param Re: Reynolds number.
:param Pr: Prandtl number.
:param rho: Fluid density.
:param u: Fluid velocity.
:return: Optimal roughness parameters.
"""
R_a, R_q, R_z = calculate_roughness_parameters(y_values, L)
eps_opt = R_q

def objective(eps):
    Nu = nusselt_number(Re, Pr, eps, D)
    delta_p = pressure_drop(L, D, rho, u, eps, Re)
    return (Nu/delta_p)

result = minimize_scalar(objective, bounds=(0, R_z),
    method='bounded')

return eps_opt, result.fun


# Sample data and function calls
y_deviations = [0.1, 0.15, 0.07, -0.02, 0.0, 0.05]
L = 10.0
D = 0.05
Re = 10000
Pr = 0.7
rho = 1.2
u = 15.0

R_a, R_q, R_z = calculate_roughness_parameters(y_deviations, L)
Nu = nusselt_number(Re, Pr, R_q, D)
delta_p = pressure_drop(L, D, rho, u, R_q, Re)
eps_opt, performance = optimize_roughness(y_deviations, L, D, Re,
    Pr, rho, u)

print(f"R_a: {R_a}, R_q: {R_q}, R_z: {R_z}")
print(f"Nusselt Number: {Nu}")
print(f"Pressure Drop: {delta_p}")
print(f"Optimal Roughness (epsilon): {eps_opt} with performance
    metric: {performance}")
```

The provided code defines and executes several key functions related to the chapter's topic:

- calculate_roughness_parameters computes statistical roughness parameters from surface height deviations.

- `nusselt_number` evaluates the convective heat transfer coefficient for turbulent flow over rough surfaces.

- `pressure_drop` uses the Darcy–Weisbach equation incorporating roughness for estimating pressure losses.

- `optimize_roughness` balances thermal performance and pressure drop, determining optimal roughness parameters.

Example data and function calls illustrate evaluating roughness effects and optimizing for desired thermal and hydraulic performance.

Chapter 45

Material Fatigue Analysis

Introduction to Fatigue in Aerospace Materials

In aerospace engineering, the structural durability of materials is critical, particularly under varying and cyclic loads encountered during launch, flight, and re-entry stages. Material fatigue, characterized by progressive degradation under repeated stress cycles, presents significant challenges in predicting the lifespan of rocket components. Rocket materials are subjected to cyclic thermal and mechanical stresses, necessitating precise fatigue analysis to ensure structural integrity and safety.

Fundamental Concepts of Material Fatigue

The fatigue phenomenon is analyzed through principles such as stress-life (S-N) curves, where the total number of cycles to failure is plotted against applied stress:

$$\log N = \log A - m \cdot \log S$$

Where N is the number of cycles to failure, S represents the stress amplitude, A is a material constant, and m is the fatigue strength exponent. The characterization of fatigue life under different loading conditions leads to identifying fatigue limits and endurance limits for specific materials.

Cyclic Load and Microstructural Effects

The analysis of cyclic loads involves the application of mean stress correction factors such as the Goodman or Gerber relation:

$$\frac{\sigma_a}{\sigma'_f} = 1 - \frac{\sigma_m}{\sigma_u}$$

Where σ_a is the stress amplitude, σ'_f is the fatigue strength in the absence of mean stress, σ_m is the mean stress, and σ_u is the ultimate tensile strength. Cyclic loading leads to microstructural changes such as slip band formation, crack initiation, and propagation, significantly affecting material fatigue behavior.

Paris' Law for Crack Growth Prediction

For crack propagation analysis under cyclic loading, Paris' Law is employed:

$$\frac{da}{dN} = C \cdot (\Delta K)^m$$

Where $\frac{da}{dN}$ is the crack growth rate per cycle, ΔK represents the stress intensity factor range, C and m are material-specific constants. This relationship highlights how incremental crack advancement correlates with fluctuating load conditions.

Miner's Rule for Cumulative Damage

Engineers often use Miner's Rule to assess cumulative damage in structures subjected to variable loading cycles:

$$D = \sum_{i=1}^{n} \frac{n_i}{N_i}$$

Here, D is the damage fraction, n_i is the number of cycles at each stress level i, and N_i is the number of cycles to failure at that stress level. The summation over all cycles provides a quantitative measure of accumulated fatigue damage, aiding in lifecycle predictions and maintenance strategies.

Computational Fatigue Analysis Techniques

Advanced computational techniques such as Finite Element Analysis (FEA) offer detailed insights into fatigue behavior. Utilizing software to model stress distribution and predict fatigue hotspots involves complex algorithms, validating theoretical equations and experimental data. The incorporation of multi-scale modeling bridges microscopic material behavior with macroscopic structural performance, enhancing predictive accuracy.

Fatigue Testing and Validation

Fatigue testing procedures, including axial, bending, and rotating beam tests, provide empirical data necessary for validating fatigue models. Statistical methods are employed to interpret test results, characterized by the mean and scatter of fatigue cycles. Establishing confidence intervals in these results ensures robust predictions of component lifespan under aerospace operational conditions.

Python Code Snippet

Below is a Python code snippet covering material fatigue analysis calculations, including stress-life curve, Paris' Law implementation, and Miner's Rule for cumulative damage.

```python
import numpy as np

def stress_life_curve(S, A, m):
    '''
    Calculate the number of cycles to failure using the stress-life
      curve.
    :param S: Stress amplitude.
    :param A: Material constant.
    :param m: Fatigue strength exponent.
    :return: Number of cycles to failure.
    '''
    return A * S**-m

def goodman_relation(sigma_a, sigma_m, sigma_f_prime, sigma_u):
    '''
    Calculate the corrected stress amplitude using the Goodman
      relation.
    :param sigma_a: Stress amplitude.
    :param sigma_m: Mean stress.
```

```
    :param sigma_f_prime: Fatigue strength in absence of mean
    ↪   stress.
    :param sigma_u: Ultimate tensile strength.
    :return: Corrected stress amplitude.
    '''
    return sigma_a / (1 - sigma_m / sigma_u)

def paris_law(da_dN, DeltaK, C, m):
    '''
    Calculate crack growth rate per cycle using Paris' Law.
    :param da_dN: Crack growth rate per cycle.
    :param DeltaK: Stress intensity factor range.
    :param C: Material-specific constant.
    :param m: Material-specific exponent.
    :return: Crack growth rate per cycle.
    '''
    return C * (DeltaK)**m

def miners_rule(n_i, N_i):
    '''
    Calculate cumulative damage using Miner's Rule.
    :param n_i: Number of cycles at each stress level.
    :param N_i: Number of cycles to failure at that stress level.
    :return: Damage fraction.
    '''
    return np.sum(n_i / N_i)

# Example usage and parameter setup
stress_levels = [150, 200, 250]   # Example stress amplitudes
cycles = [1e5, 5e4, 2e4]    # Example cycles to failure for each
↪   stress level
n_cycles = [30000, 20000, 10000]   # Number of cycles experienced at
↪   each stress level

# Parameters for Goodman relation
sigma_a = 200
sigma_m = 50
sigma_f_prime = 250
sigma_u = 500

# Parameters for Paris' Law
DeltaK = 35
crack_growth_rate = 0.05
C = 1e-10
m_paris = 3.0

# Calculations
num_cycles_to_failure = stress_life_curve(stress_levels, A=1e12,
↪   m=5)
corrected_stress_amplitude = goodman_relation(sigma_a, sigma_m,
↪   sigma_f_prime, sigma_u)
paris_crack_growth = paris_law(crack_growth_rate, DeltaK, C,
↪   m_paris)
```

```
damage_fraction = miners_rule(n_cycles, cycles)

print("Number of cycles to failure:", num_cycles_to_failure)
print("Corrected stress amplitude:", corrected_stress_amplitude)
print("Paris crack growth rate:", paris_crack_growth)
print("Cumulative damage fraction:", damage_fraction)
```

This code outlines key functions for material fatigue analysis in aerospace engineering:

- `stress_life_curve` calculates the number of cycles to failure using the stress-life (S-N) curve methodology.

- `goodman_relation` uses a mean stress correction factor, demonstrating its application in adjusting stress amplitudes for fatigue calculations.

- `paris_law` estimates the rate of crack growth per cycle, essential for predictive maintenance and safety assurance using Paris' Law.

- `miners_rule` aggregates damage from multiple stress levels using Miner's Rule to evaluate the remaining life of components.

The provided example demonstrates utilizing these computations for assessing structural integrity and lifespan of materials under cyclic stresses.

Chapter 46

Supersonic Flow Calculations

Introduction to Supersonic Flow

Supersonic flow around rockets is characterized by velocities exceeding the speed of sound, resulting in distinct physical phenomena such as shock waves and rapid changes in flow properties. The flow regimes in these conditions necessitate advanced analytical techniques to account for compressibility effects and thermal variations. These phenomena significantly influence aerodynamic forces, especially drag, necessitating precise calculations to ensure optimal performance and structural integrity of rockets during supersonic flight.

Fundamental Equations of Supersonic Flow

The behavior of supersonic flow can be fundamentally described by the compressible flow equations, where the changes in flow properties are governed by:

$$\frac{dp}{p} = \gamma M^2 \frac{d\rho}{\rho} = -\gamma M^2 \frac{dV}{V}$$

Here, p denotes pressure, ρ represents density, V is velocity, γ is the specific heat ratio, and M is the Mach number. At supersonic speeds, the relationship between these parameters becomes critical in evaluating pressure and density variations across shock waves and expansion fans.

Shock Wave Formation

Shock waves are a defining feature of supersonic flow, manifesting as abrupt, discontinuous changes in flow properties. The classical normal shock wave relations for a perfect gas are derived from the conservation of mass, momentum, and energy, expressed as:

$$\frac{p_2}{p_1} = 1 + \frac{2\gamma}{\gamma+1}(M_1^2 - 1)$$

$$\frac{\rho_2}{\rho_1} = \frac{(\gamma+1)M_1^2}{2 + (\gamma-1)M_1^2}$$

$$\frac{T_2}{T_1} = \frac{(2\gamma M_1^2 - (\gamma-1))((\gamma-1)M_1^2 + 2)}{(\gamma+1)^2 M_1^2}$$

where the subscripts 1 and 2 represent flow properties upstream and downstream of the shock. These equations provide a mathematical framework to analyze shock strength and its impact on flow characteristics.

Drag Reduction Techniques

Reducing drag in supersonic flow contributes significantly to rocket performance. One approach involves optimizing the body shape to mitigate shockwave interactions and boundary layer separations. The drag force D on a body can be broadly decomposed as:

$$D = D_{\text{pressure}} + D_{\text{friction}}$$

Drag reduction techniques emphasize minimizing D_{pressure} by controlling expansion and compression patterns. Strategy involves using supersonic area rule designs and controlling flow separation through aerodynamic surfaces, contributing to minimized wave drag and enhanced stability.

Numerical Methods for Supersonic Flow

Supersonic flow calculations incorporate numerical methods for resolving the complex, non-linear partial differential equations governing flow dynamics. The computational fluid dynamics (CFD) approach deals with the Euler and Navier-Stokes equations in discretized forms:

$$\frac{\partial \mathbf{U}}{\partial t} + \frac{\partial \mathbf{F(U)}}{\partial x} + \frac{\partial \mathbf{G(U)}}{\partial y} + \frac{\partial \mathbf{H(U)}}{\partial z} = 0$$

where $\mathbf{U}$ refers to the conserved variable vector, and $\mathbf{F}, \mathbf{G}, \mathbf{H}$ denote the flux vectors. Advanced turbulence models and high-resolution schemes are employed to simulate shockwave interactions and boundary conditions accurately.

Applications in Modern Rocket Design

In modern aerospace applications, understanding supersonic flow dynamics guides critical design decisions, impacting factors such as stage separation, nozzle design, and thermal control systems. Detailed supersonic flow assessments ensure that rockets maintain stability, minimize drag, and manage thermal loads effectively under flight conditions traversing multiple Mach regimes.

Python Code Snippet

Below is a Python code snippet that encompasses the core computational elements of supersonic flow calculations including fundamental equations and numerical simulation methods.

```python
import numpy as np

def fundamental_supersonic_flow(p, rho, V, gamma, M):
    '''
    Calculate fundamental changes in supersonic flow properties.
    :param p: Pressure.
    :param rho: Density.
    :param V: Velocity.
    :param gamma: Specific heat ratio.
    :param M: Mach number.
    :return: Changes in pressure and density.
    '''
    dp_over_p = gamma * M**2 * (1/rho)
```

```
        drho_over_rho = -gamma * M**2 * (1/V)
        return dp_over_p, drho_over_rho

def normal_shock_relations(M1, gamma):
    '''
    Calculate the properties across a normal shock wave.
    :param M1: Upstream Mach number.
    :param gamma: Specific heat ratio.
    :return: Ratios p2/p1, rho2/rho1, T2/T1.
    '''
    p_ratio = 1 + (2 * gamma / (gamma + 1)) * (M1**2 - 1)
    rho_ratio = ((gamma + 1) * M1**2) / (2 + (gamma - 1) * M1**2)
    T_ratio = ((2 * gamma * M1**2 - (gamma - 1)) * ((gamma - 1) *
    ↪   M1**2 + 2)
                / ((gamma + 1)**2 * M1**2))
    return p_ratio, rho_ratio, T_ratio

def drag_decomposition(D_pressure, D_friction):
    '''
    Decompose total drag into pressure and friction components.
    :param D_pressure: Pressure drag component.
    :param D_friction: Friction drag component.
    :return: Total drag.
    '''
    return D_pressure + D_friction

def supersonic_cfd_simulation(U, F, G, H, dx, dy, dz, dt):
    '''
    Perform numerical simulation of supersonic flow using Euler
    ↪   equations.
    :param U: Conserved variable vector.
    :param F, G, H: Flux vectors.
    :param dx, dy, dz: Spatial step sizes.
    :param dt: Time step.
    :return: Updated conserved variables after one timestep.
    '''
    dU_dt = - ((F(U) / dx) + (G(U) / dy) + (H(U) / dz))
    U_new = U + dt * dU_dt
    return U_new

def calculate_flow(Mach, gamma=1.4):
    '''
    Example function to encapsulate supersonic flow calculations.
    :param Mach: Mach number.
    :param gamma: Specific heat ratio, default is 1.4 for air.
    :return: Properties ratios across shock and total drag.
    '''
    # Initial parameters (example values)
    p = 101325   # Pressure in Pascals
    rho = 1.225  # Density in kg/m^3
    V = 340      # Velocity in m/s
    D_pressure = 100  # Example pressure drag
    D_friction = 50   # Example friction drag
```

```python
# Fundamental supersonic flow calculations
dp_over_p, drho_over_rho = fundamental_supersonic_flow(p, rho,
    V, gamma, Mach)

# Calculate shock wave relations
p_ratio, rho_ratio, T_ratio = normal_shock_relations(Mach,
    gamma)

# Total drag
total_drag = drag_decomposition(D_pressure, D_friction)

return {
    'dp_over_p': dp_over_p,
    'drho_over_rho': drho_over_rho,
    'shock_p_ratio': p_ratio,
    'shock_rho_ratio': rho_ratio,
    'shock_T_ratio': T_ratio,
    'total_drag': total_drag
}

# Example usage of calculate_flow function
s_flow_properties = calculate_flow(2.0)
for key, value in s_flow_properties.items():
    print(f"{key}: {value}")
```

This code defines several key functions necessary for understanding and simulating supersonic flow characteristics:

- `fundamental_supersonic_flow` calculates changes in flow properties that account for compressibility effects.

- `normal_shock_relations` computes the properties across a normal shock wave using upstream Mach numbers.

- `drag_decomposition` breaks down the total drag into pressure and friction components.

- `supersonic_cfd_simulation` represents an Euler equation based numerical simulation framework for studying flow dynamics.

- `calculate_flow` provides an encapsulation of supersonic flow calculations including shock properties and drag assessment, integrating individual computations.

This Python script serves as a fundamental toolset for engineers and researchers involved in analyzing and optimizing supersonic rocket designs.

Chapter 47

Flow Separation and Boundary Layer Theory

Boundary Layer Formation in High-Speed Flow

The boundary layer represents a region adjacent to a solid boundary where viscous effects are significant. In high-speed flows, such as those encountered in rocket configurations, the dynamics of boundary layer development profoundly impact aerodynamic performance. The boundary layer thickness, δ, can be estimated using:

$$\delta \approx 5\sqrt{\frac{\nu x}{U_\infty}}$$

where ν is the kinematic viscosity, x is the distance from the leading edge, and U_∞ is the free-stream velocity. The interplay between inertial and viscous forces governs the boundary layer characteristics.

Turbulent Boundary Layer Considerations

In high-speed rocket configurations, transition to turbulence often occurs within the boundary layer, significantly affecting drag and heat transfer rates. The turbulent boundary layer is characterized by complex eddy motion, necessitating a different approach to modeling compared to laminar flow. The Reynolds-averaging approach to the Navier-Stokes equations introduces additional terms, the Reynolds stresses:

$$-\overline{\rho u'v'} = \mu_t \frac{\partial \overline{U}}{\partial y}$$

where μ_t is the eddy viscosity, representing the enhanced mixing due to turbulence. Turbulence models, such as the $k - \epsilon$ and $k - \omega$ models, provide closure to the Reynolds-averaged Navier-Stokes equations.

Flow Separation Mechanisms

Flow separation occurs when the boundary layer detaches from the surface of the rocket, often resulting from adverse pressure gradients. This phenomenon causes a significant increase in drag and loss of lift. The point of separation can be theoretically predicted using the Clauser parameter, λ:

$$\lambda = \frac{\beta \delta^*}{\theta}$$

where β is the pressure gradient parameter, δ^* is the displacement thickness, and θ is the momentum thickness. Critical values for λ correlate with onset of separation.

Calculations of Displacement and Momentum Thickness

Quantifying boundary layer characteristics involves calculating both displacement thickness, δ^*, and momentum thickness, θ. These are defined respectively as:

$$\delta^* = \int_0^\infty \left(1 - \frac{U}{U_\infty}\right) dy$$

$$\theta = \int_0^\infty \frac{U}{U_\infty} \left(1 - \frac{U}{U_\infty}\right) dy$$

where U is the local velocity within the boundary layer. These integral quantities are critical in determining the effective boundary layer thickness contributing to flow separation.

Impact of Compressibility on Boundary Layer Development

Compressibility effects become significant in high-speed flows, altering the characteristics of the boundary layer. The compressible boundary layer thickness differs from the incompressible one and can be corrected using the Van Driest transformation:

$$y^+ = \frac{yU_\tau}{\nu} \quad \text{and} \quad U^+ = \int_0^{U/U_\tau} \sqrt{\frac{\rho}{\rho_w}} \, d\left(\frac{U}{U_\tau}\right)$$

where U_τ is the friction velocity, and the subscript w denotes properties at the wall. These corrections are essential for accurate prediction and analysis.

Mitigation Strategies for Flow Separation

Various strategies exist to delay or mitigate flow separation in high-speed applications. Techniques such as boundary layer suction, vortex generators, and careful contouring of the rocket surfaces are employed. Numerical simulations often explore these approaches using Computational Fluid Dynamics (CFD), where boundary layer development and separation points are precisely monitored using appropriate turbulence models.

These sections cohesively describe the complex mechanisms governing flow separation and boundary layer theory in the context of high-speed rocket applications, vital for optimizing aerodynamic performance and stability.

Python Code Snippet

Below is a Python code snippet that encompasses the core computational elements discussed in the chapter, including boundary layer thickness calculations, turbulent flow modeling, and flow separation dynamics.

```python
import numpy as np
import scipy.integrate as spi

# Define constants
nu = 1.5e-5  # kinematic viscosity (m^2/s)
U_inf = 100  # free-stream velocity (m/s)
x = 1.0  # distance from leading edge (m)

def boundary_layer_thickness(nu, x, U_inf):
    '''
    Calculate the boundary layer thickness.
    :param nu: Kinematic viscosity.
    :param x: Distance from the leading edge.
    :param U_inf: Free-stream velocity.
    :return: Boundary layer thickness.
    '''
    delta = 5 * np.sqrt(nu * x / U_inf)
    return delta

# Example calculation for boundary layer thickness
delta = boundary_layer_thickness(nu, x, U_inf)

def reynolds_stress(U, U_inf, y, mu_t):
    '''
    Calculate the Reynolds stress using a turbulence model.
    :param U: Local velocity in the boundary layer.
    :param U_inf: Free-stream velocity.
    :param y: Distance from the wall.
    :param mu_t: Eddy viscosity.
    :return: Reynolds stress.
    '''
    dU_dy = np.gradient(U, y)  # velocity gradient
    tau = -mu_t * dU_dy
    return tau

def displacement_thickness(U, U_inf, y):
    '''
    Calculate the displacement thickness.
    :param U: Local velocity in the boundary layer.
    :param U_inf: Free-stream velocity.
    :param y: Distance from the wall.
    :return: Displacement thickness.
    '''
    integrand = lambda y: (1 - U(y) / U_inf)
    delta_star = spi.quad(integrand, 0, np.inf)[0]
    return delta_star

def momentum_thickness(U, U_inf, y):
    '''
    Calculate the momentum thickness.
    :param U: Local velocity in the boundary layer.
    :param U_inf: Free-stream velocity.
```

```
    :param y: Distance from the wall.
    :return: Momentum thickness.
    '''
    integrand = lambda y: (U(y) / U_inf) * (1 - U(y) / U_inf)
    theta = spi.quad(integrand, 0, np.inf)[0]
    return theta

def flow_separation_criteria(delta_star, theta, beta):
    '''
    Predict flow separation using the Clauser parameter.
    :param delta_star: Displacement thickness.
    :param theta: Momentum thickness.
    :param beta: Pressure gradient parameter.
    :return: Clauser parameter for separation prediction.
    '''
    lambda_clauser = (beta * delta_star) / theta
    return lambda_clauser

# Example usage of flow separation criteria
U = lambda y: U_inf * (1 - np.exp(-0.1 * y))  # Example velocity
↪ profile
y = np.linspace(0, 0.5, 100)  # Distance from wall
mu_t = 1.0e-4  # Eddy viscosity (example value)

reynolds_stress(U, U_inf, y, mu_t)  # Example Reynolds stress
↪ calculation
delta_star = displacement_thickness(U, U_inf, y)
theta = momentum_thickness(U, U_inf, y)
beta = 0.9  # Pressure gradient parameter (example value)

lambda_clauser = flow_separation_criteria(delta_star, theta, beta)

print("Boundary layer thickness:", delta)
print("Displacement thickness:", delta_star)
print("Momentum thickness:", theta)
print("Clauser parameter:", lambda_clauser)
```

This code defines several key functions necessary for the analysis of boundary layer flow:

- `boundary_layer_thickness` function computes the boundary layer thickness based on the kinematic viscosity, distance from the leading edge, and free-stream velocity.

- `reynolds_stress` calculates the Reynolds stress using a turbulence model, taking into account the local velocity gradient and eddy viscosity.

- `displacement_thickness` evaluates the displacement thickness using numerical integration over the velocity profile.

- `momentum_thickness` computes the momentum thickness similarly by integrating over the boundary layer.

- `flow_separation_criteria` uses the Clauser parameter to predict the onset of flow separation, which is key for optimizing aerodynamic performance.

The final block of code provides examples of using these functions with example data.

Chapter 48

Free-Molecular Flow and its Effects

Regime Characteristics of Free-Molecular Flow

Free-molecular flow emerges when a vehicle operates at exceedingly high altitudes where the mean free path l becomes comparable to the characteristic length of the object L, typically characterized by the Knudsen number $\mathrm{Kn} = \frac{l}{L} \gg 1$. Under these conditions, the continuum hypothesis breaks down, and molecular interactions with surfaces become predominant. The gas molecules travel in ballistic trajectories between collisions with the vehicle surface, influenced by molecular interaction models.

Mathematical Modeling of Free-Molecular Flow

In contrast to continuum flow regimes, the mathematical representation of free-molecular flow lacks the applicability of the Navier-Stokes equations, necessitating statistical mechanics and kinetic theory. The velocity distribution function $f(\mathbf{v}, \mathbf{x}, t)$ describes the number density of molecules at position $\mathbf{x}$ and time t with velocity $\mathbf{v}$.

The Boltzmann equation, simplified in the absence of intermolecular collisions, governs the evolution of f:

$$\frac{\partial f}{\partial t} + \mathbf{v} \cdot \nabla_{\mathbf{x}} f = 0$$

Boundary interactions on the vehicle surface influence f, characterized by diffuse reflection models where the post-collision molecular velocities are distributed according to a Maxwellian distribution.

Drag Force in Free-Molecular Flow

The drag force in this regime, F_D, is predominantly a result of momentum exchange between the molecules and the vehicle surface. The analytical prediction of F_D is derived by integrating molecular impingement dynamics over the surface:

$$F_D = \int_{\text{surface}} \int_{\text{velocity space}} m(\mathbf{v} \cdot \mathbf{n})(\mathbf{v} - \mathbf{v}_w) f(\mathbf{v}) \, d\mathbf{v} \, dA$$

where m is the molecular mass, $\mathbf{v}_w$ is the velocity of the wall, and $\mathbf{n}$ is the unit normal to the surface.

Thermal Effects and Energy Transfer

Energy transfer in free-molecular flow involves the accommodation coefficient α, which represents the fraction of kinetic energy exchanged during molecule-surface encounters. For thermal analysis, the heat flux q to the surface is given by:

$$q = \alpha \int_{\text{surface}} \int_{\text{velocity space}} \frac{1}{2} m (\mathbf{v} \cdot \mathbf{n})(v^2 - v_w^2) f(\mathbf{v}) \, d\mathbf{v} \, dA$$

Understanding this energy exchange is critical for thermal protection system design in high-altitude vehicle configurations.

Predicting Vehicle Performance in Free-Molecular Flow

Vehicle performance predictions under free-molecular conditions depend significantly on the vehicle's geometric and surface characteristics. The analysis of the translational, rotational, and vibrational energy modes of molecules inform the vehicle's momentum and energy exchange processes. Computational approaches often involve Monte Carlo simulation techniques to statistically model molecular interactions and their resulting force and heat transfer distributions across the vehicle surface.

The influences of geometry, including sharp leading edges and flat surfaces, are critical in these predictions. The complexity of these interactions requires comprehensive modeling to ensure structural integrity and operational efficiency in high-altitude aerospace applications, where free-molecular flow conditions predominate.

Python Code Snippet

Below is a Python code snippet that encompasses the core computational elements associated with free-molecular flow, including drag force calculations and thermal effects in high-altitude vehicle applications.

```
import numpy as np

def boltzmann_velocity_distribution(v, v_w, n, m, T):
    '''
    Calculate the velocity distribution function for free-molecular
    ↪ flow.
    :param v: Molecular velocity.
    :param v_w: Wall velocity.
    :param n: Number density of molecules.
    :param m: Mass of molecules.
    :param T: Temperature.
    :return: Velocity distribution.
    '''
    k = 1.38e-23  # Boltzmann constant
    return n * ((m / (2 * np.pi * k * T)) ** 1.5) * np.exp(-m * (v -
    ↪ v_w)**2 / (2 * k * T))

def drag_force_on_surface(surface_area,
↪ velocity_distribution_function, v_w, n_vector,
                          m_array):
    '''
    Calculate the drag force on the vehicle surface.
    :param surface_area: Surface area of the vehicle.
    :param velocity_distribution_function: The velocity distribution
    ↪ function.
    :param v_w: Wall velocity.
    :param n_vector: Surface normal vector.
    :param m_array: Molecular mass array.
    :return: Drag force.
    '''
```

```
    F_D = 0.0

    for v, m in zip(np.linspace(-500, 500, num=100), m_array):
        f_v = velocity_distribution_function(v, v_w, 1e19, m, 300)
          # use dummy values for n, T
        F_D += m * np.dot(v, n_vector) * (v - v_w) * f_v *
          surface_area

    return F_D

def heat_flux_to_surface(surface_area,
    velocity_distribution_function,
                         v_w, n_vector, m_array, alpha):
    '''
    Calculate the heat flux to the surface.
    :param surface_area: Surface area of the vehicle.
    :param velocity_distribution_function: The velocity distribution
      function.
    :param v_w: Wall velocity.
    :param n_vector: Surface normal vector.
    :param m_array: Molecular mass array.
    :param alpha: Accommodation coefficient.
    :return: Heat flux.
    '''
    q = 0.0

    for v, m in zip(np.linspace(-500, 500, num=100), m_array):
        f_v = velocity_distribution_function(v, v_w, 1e19, m, 300)
          # use dummy values for n, T
        kinetic_energy = 0.5 * m * (v**2 - v_w**2)
        q += alpha * kinetic_energy * np.dot(v, n_vector) * f_v *
          surface_area

    return q

# Example of evaluating these calculations
molecular_mass = np.array([4.65e-26, 2.66e-26])  # masses for
  example molecules in kg

surface_normal = np.array([0, 1, 0])  # assume surface normal is
  along y-axis
vehicle_area = 1.0  # dummy area in m^2
vehicle_velocity = np.array([0, 500, 0])  # wall velocity in m/s
  along y-axis

drag = drag_force_on_surface(vehicle_area,
    boltzmann_velocity_distribution,
                             vehicle_velocity, surface_normal,
                               molecular_mass)
heat_flux = heat_flux_to_surface(vehicle_area,
    boltzmann_velocity_distribution,
                                 vehicle_velocity, surface_normal,
                                   molecular_mass, 0.9)
```

```
print("Drag Force:", drag)
print("Heat Flux:", heat_flux)
```

This code defines several key functions necessary for modeling free-molecular flow effects on high-altitude vehicles:

- `boltzmann_velocity_distribution`: Computes the velocity distribution function based on Boltzmann's formula.

- `drag_force_on_surface`: Calculates the drag force exerted by high-altitude molecular impacts on the vehicle's surface.

- `heat_flux_to_surface`: Computes the heat flux transferred to the vehicle's surface from molecular impacts, accounting for energy exchanges.

The final segment of code provides examples using placeholder values to demonstrate drag force and heat flux evaluations within the free-molecular flow regime.

Chapter 49

Cryogenic Propellant Management

Introduction to Cryogenic Propellants

Cryogenic propellants, characterized by their ultra-low temperatures, require meticulous management within aerospace systems to maintain efficacy and safety throughout the mission profile. Key cryogens such as Liquid Hydrogen (LH_2) and Liquid Oxygen (LO_2) exhibit unique properties that necessitate precise thermodynamic and fluid dynamic calculations to ensure their stable storage and handling.

Thermodynamic Properties of Cryogenic Fluids

The thermodynamic behavior of cryogenic fluids is governed by their phase diagrams, encapsulating equilibrium between liquid and vapor phases. The Clapeyron equation,

$$\frac{dP}{dT} = \frac{L}{T\Delta v}$$

where L is the latent heat of vaporization, T is temperature, and Δv is the change in specific volume, delineates phase transition phenomena critically relevant at the cryogenic temperature range.

Heat Transfer Mechanisms in Cryogenic Storage

Effective insulation against thermal influx is paramount for minimizing cryogen boil-off. Convection, conduction, and radiation are the primary heat transfer modes affecting cryogenic tanks. The heat transfer rate $\dot{Q}$ through multilayer thermal insulation can be approximated as:

$$\dot{Q} = \frac{kA(T_1 - T_2)}{d}$$

where k is the thermal conductivity, A is the surface area, T_1 and T_2 are the boundary temperatures, and d is the insulation thickness.

Cryogenic Fluid Dynamics

Cryogenic fluid motion bears similarity to incompressible flows, albeit with significant temperature dependence on viscosity and density. Utilizing the Navier-Stokes equations, the velocity profile $u(x, y, z, t)$ is determined by:

$$\rho \left(\frac{\partial \mathbf{u}}{\partial t} + (\mathbf{u} \cdot \nabla)\mathbf{u} \right) = -\nabla P + \mu \nabla^2 \mathbf{u} + \rho \mathbf{g}$$

where ρ is fluid density, P is pressure, μ is dynamic viscosity, and $\mathbf{g}$ is gravitational acceleration.

Boil-off and Phase Change Calculations

Evaluating boil-off rates necessitates an integration of energy balance principles. The boil-off rate $\dot{m}$ is assessed using:

$$\dot{m} = \frac{\dot{Q}}{L_v}$$

where L_v represents the latent heat of vaporization, critical in designing cryogenic storage systems with minimal product loss.

Storage Tank Design Considerations

The structural integrity of cryogenic tanks involves analyzing stress distributions caused by low-temperature contractions. Using finite element analysis (FEA), the stress σ is calculated as:

$$\sigma = \frac{F}{A}$$

where F is the applied force and A is the cross-sectional area. Material selectivity is predicated on maintaining ductility at cryogenic temperatures, requiring alloys with low temperature coefficients.

Conclusion

Through rigorous thermodynamic, fluid dynamic, and structural principles, cryogenic propellant management within aerospace applications is enhanced, optimizing storage longevity and system performance during mission operations.

Python Code Snippet

Below is a Python code snippet that includes key calculations and algorithms related to cryogenic propellant management, such as heat transfer, fluid dynamics, and tank design.

```
import numpy as np

def clapeyron_equation(L, T, delta_v):
    """
    Calculate the change in pressure with temperature using the
      Clapeyron equation.
    :param L: Latent heat of vaporization.
    :param T: Temperature in Kelvin.
    :param delta_v: Change in specific volume.
    :return: Change in pressure.
    """
    return L / (T * delta_v)

def heat_transfer_rate(k, A, T1, T2, d):
    """
    Calculate the heat transfer rate through insulation.
    :param k: Thermal conductivity.
    :param A: Surface area.
```

```
    :param T1: Temperature at the outer boundary.
    :param T2: Temperature at the inner boundary.
    :param d: Insulation thickness.
    :return: Heat transfer rate.
    """
    return (k * A * (T1 - T2)) / d

def navier_stokes_density_velocity(rho, u, P, mu, g):
    """
    Calculate velocity profile using the Navier-Stokes equation.
    :param rho: Density of the fluid.
    :param u: Velocity vector.
    :param P: Pressure.
    :param mu: Dynamic viscosity.
    :param g: Gravitational acceleration vector.
    :return: Velocity profile.
    """
    # Here we would perform numerical integration, using
    ↪   placeholders for simplicity
    velocity_profile = rho * (np.gradient(u) + np.cross(u,
    ↪   np.gradient(u))) - np.gradient(P) + mu *
    ↪   np.gradient(np.gradient(u)) + rho * g
    return velocity_profile

def boil_off_rate(Q_dot, Lv):
    """
    Calculate the boil-off rate of the cryogenic fluid.
    :param Q_dot: Heat transfer rate.
    :param Lv: Latent heat of vaporization.
    :return: Boil-off rate.
    """
    return Q_dot / Lv

def stress_distribution(F, A):
    """
    Calculate stress within storage tank material.
    :param F: Force applied.
    :param A: Cross-sectional area.
    :return: Stress.
    """
    return F / A

# Example computations
latent_heat = 2260  # J/kg
temperature = 20 + 273.15  # K
specific_volume_change = 1.0  # m^3/kg
pressure_change = clapeyron_equation(latent_heat, temperature,
↪   specific_volume_change)

thermal_conductivity = 0.1  # W/(m*K)
surface_area = 10.0  # m^2
outer_temperature = 300.0  # K
inner_temperature = 77.0  # K
```

```
insulation_thickness = 0.05   # m
heat_transfer = heat_transfer_rate(thermal_conductivity,
↪   surface_area, outer_temperature, inner_temperature,
↪   insulation_thickness)

density = 71   # kg/m^3
velocity = np.array([0, 0, 0])   # Placeholder velocity vector
pressure = 101325   # Pressure in Pascals
dynamic_viscosity = 1.8e-5   # Pa.s
gravity = np.array([0, -9.81, 0])
velocity_profile = navier_stokes_density_velocity(density, velocity,
↪   pressure, dynamic_viscosity, gravity)

heat_transfer_rate_value = 500   # W, example value
latent_heat_vaporization = 442000   # J/kg, for a specific cryogen
boiloff = boil_off_rate(heat_transfer_rate_value,
↪   latent_heat_vaporization)

force = 1000   # N
cross_section_area = 0.02   # m^2
stress = stress_distribution(force, cross_section_area)

print("Change in Pressure:", pressure_change)
print("Heat Transfer Rate:", heat_transfer)
print("Velocity Profile:", velocity_profile)
print("Boil-off Rate:", boiloff)
print("Stress Distribution:", stress)
```

This code defines several key functions necessary for managing cryogenic propellants:

- `clapeyron_equation` calculates the change in pressure with temperature using the Clapeyron equation.

- `heat_transfer_rate` computes heat transfer through insulation for cryogenic tanks.

- `navier_stokes_density_velocity` determines the velocity profile within the cryogenic fluid using the Navier-Stokes equation.

- `boil_off_rate` evaluates the rate of boil-off or vapor loss from the fluid.

- `stress_distribution` calculates stress within materials used in cryogenic storage tanks.

The final block of code provides example calculations related to cryogenic propellant management using realistic physical parameters.

Chapter 50

Forced and Natural Convection in Rockets

Fundamentals of Convection Mechanisms

In the domain of aerospace propulsion, the thermal regulation of rocket components is paramount to sustaining operational integrity during ascent and re-entry phases. Convection, occurring through forced or natural mechanisms, serves as a pivotal heat transfer mode, attributable to the movement of fluid layers driven by external forces or buoyancy effects. The analytical approach to understanding these phenomena begins with the generalized convective heat transfer equation:

$$q = hA(T_s - T_\infty)$$

where q denotes the convective heat transfer rate, h the convective heat transfer coefficient, A the surface area, T_s the surface temperature, and T_∞ the fluid's bulk temperature.

Mathematical Modeling of Forced Convection

Forced convection arises when an external agent induces fluid motion over a surface, enhancing heat transport away from rocket structures. The forced convection heat transfer velocity profile along the boundary layer of a surface is governed by the dimensionless Reynolds number, Re, defined as:

$$\text{Re} = \frac{\rho U L}{\mu}$$

where ρ represents fluid density, U the characteristic flow velocity, L the characteristic length, and μ the dynamic viscosity. Using this parameter, the Nusselt number, Nu, which correlates convective to conductive heat transfer, can be expressed for turbulent flow:

$$\text{Nu} = C \cdot \text{Re}^m \cdot \text{Pr}^n$$

where C, m, and n are empirical coefficients derived from experimental calibration, and Pr is the Prandtl number given by:

$$\text{Pr} = \frac{c_p \mu}{k}$$

here, c_p denotes the specific heat capacity at constant pressure and k the thermal conductivity.

Natural Convection in Rocket Systems

Natural convection is driven by buoyant forces arising from temperature gradients within the fluid. The Grashof number, Gr, quantifies the relative significance of buoyancy to viscous forces and is defined as:

$$\text{Gr} = \frac{g\beta(T_s - T_\infty)L^3}{\nu^2}$$

where g is the acceleration due to gravity, β the thermal expansion coefficient, and ν the kinematic viscosity. For laminar flow conditions, the relationship between the Nusselt number and the Grashof and Prandtl numbers is often expressed as:

$$\text{Nu} = (\text{Gr} \cdot \text{Pr})^n$$

This relationship facilitates the prediction of heat exchange efficiency in passive cooling scenarios encountered by rockets.

Applications in Rocket Thermal Management

In the engineering practice of rocket thermal management, the balance of forced and natural convection is critical to achieving thermal equilibrium. The incorporation of computational fluid dynamics (CFD) simulations, equipped with well-posed boundary conditions, provides insights into the spatial distribution of temperature and velocity fields. Governing equations—specifically, the Navier-Stokes and energy equations—are solved numerically to predict flow patterns and thermal exchanges, informing design decisions regarding material selection and protective coatings.

CFD codes implement iterations of the discretized form of the energy equation:

$$\rho c_p \left(\frac{\partial T}{\partial t} + \mathbf{u} \cdot \nabla T \right) = k \nabla^2 T + \dot{q}_V$$

where $\dot{q}_V$ is the volumetric heat generation rate. The iterative computation ensures that the thermal loads during various flight phases do not surpass the material limits, preserving the structural integrity and functionality of critical components.

Significance of Convection Studies in Rocketry

The capability to accurately model and control convection processes ensures the efficient dissipation of heat generated by engine operations or atmospheric interactions. Through empirically derived correlations and advanced simulation tools, aerospace engineers design thermal management systems that optimize the balance between heat rejection and retention, thus maintaining desired temperature profiles across different mission segments.

Python Code Snippet

Below is a Python code snippet that captures the essential calculations related to forced and natural convection in rockets, including determining various dimensionless numbers and computing heat transfer efficiencies.

```python
import numpy as np

def compute_reynolds_number(rho, U, L, mu):
    '''
    Calculate the Reynolds number for forced convection.
    :param rho: Fluid density.
    :param U: Characteristic flow velocity.
    :param L: Characteristic length.
    :param mu: Dynamic viscosity.
    :return: Reynolds number.
    '''
    return (rho * U * L) / mu

def compute_prandtl_number(cp, mu, k):
    '''
    Calculate the Prandtl number.
    :param cp: Specific heat capacity at constant pressure.
    :param mu: Dynamic viscosity.
    :param k: Thermal conductivity.
    :return: Prandtl number.
    '''
    return (cp * mu) / k

def compute_nusselt_number_forced(Re, Pr, C=0.023, m=0.8, n=0.3):
    '''
    Calculate the Nusselt number for forced convection.
    :param Re: Reynolds number.
    :param Pr: Prandtl number.
    :param C: Empirical coefficient, default for turbulent flow.
    :param m: Empirical exponent, default for turbulent flow.
    :param n: Empirical exponent, default for Prandtl dependence.
    :return: Nusselt number.
    '''
    return C * (Re ** m) * (Pr ** n)

def compute_grashof_number(g, beta, delta_T, L, nu):
    '''
    Calculate the Grashof number for natural convection.
    :param g: Acceleration due to gravity.
    :param beta: Thermal expansion coefficient.
    :param delta_T: Temperature difference (T_s - T_infinity).
    :param L: Characteristic length.
    :param nu: Kinematic viscosity.
    :return: Grashof number.
    '''
    return (g * beta * delta_T * L ** 3) / (nu ** 2)

def compute_nusselt_number_natural(Gr, Pr, n=1/3):
    '''
    Calculate the Nusselt number for natural convection.
    :param Gr: Grashof number.
    :param Pr: Prandtl number.
```

```
    :param n: Empirical exponent for GrPr product.
    :return: Nusselt number.
    '''
    return (Gr * Pr) ** n

# Example Parameters
rho = 1.225    # Air density at sea level (kg/m^3)
U = 10.0       # Flow velocity (m/s)
L = 0.5        # Characteristic length (m)
mu = 1.81e-5   # Dynamic viscosity of air (Pa.s)
cp = 1005      # Specific heat capacity at constant pressure (J/(kg.K))
k = 0.0257     # Thermal conductivity of air (W/(m.K))
g = 9.81       # Gravity (m/s^2)
beta = 1/300   # Approx. thermal expansion coefficient (1/K)
delta_T = 50   # Temperature difference (K)
nu = mu / rho  # Kinematic viscosity (m^2/s)

# Calculations
Re = compute_reynolds_number(rho, U, L, mu)
Pr = compute_prandtl_number(cp, mu, k)
Nu_forced = compute_nusselt_number_forced(Re, Pr)
Gr = compute_grashof_number(g, beta, delta_T, L, nu)
Nu_natural = compute_nusselt_number_natural(Gr, Pr)

print("Reynolds Number:", Re)
print("Prandtl Number:", Pr)
print("Nusselt Number (Forced):", Nu_forced)
print("Grashof Number:", Gr)
print("Nusselt Number (Natural):", Nu_natural)
```

This code offers functions to calculate key dimensionless parameters used to assess heat transfer in convection scenarios:

- `compute_reynolds_number` function evaluates the Reynolds number, indicating the flow regime.

- `compute_prandtl_number` calculates the Prandtl number, linking momentum diffusivity to thermal diffusivity.

- `compute_nusselt_number_forced` determines the Nusselt number for forced convection.

- `compute_grashof_number` calculates the Grashof number, important for assessing natural convection effects.

- `compute_nusselt_number_natural` evaluates the Nusselt number for natural convection scenarios.

The example calculations with typical air properties demonstrate the utility of these functions in practical applications concerning thermal management in rockets.

Chapter 51

Failure Mode and Effect Analysis (FMEA)

Introduction to FMEA in Propulsion Systems

Failure Mode and Effect Analysis (FMEA) is a structured approach used to identify and mitigate potential failure modes in propulsion systems. The methodology seeks to enhance system reliability by systematically analyzing each component's potential failure points and their subsequent effects on the overall system. FMEA allows for the prioritization of risks, directing focus toward the most critical areas requiring engineering attention.

FMEA Methodology

The FMEA process begins with the identification of potential failure modes across various components of the propulsion system. For each potential failure mode, the methodology assesses the severity (S), occurrence (O), and detection (D) metrics. These metrics are compiled into a Risk Priority Number (RPN), calculated as:

$$\text{RPN} = S \times O \times D$$

The RPN aids in prioritizing failure modes based on their potential impact, likelihood of occurrence, and ease of detection.

1 Severity Assessment

Severity (S) is a measure of the consequences of a failure mode. The severity level depends on the failure's potential impact on system operations, safety, and mission success. A higher severity rating is assigned to failure modes with more detrimental effects on the propulsion system's performance.

2 Occurrence Evaluation

Occurrence (O) quantifies the likelihood of each failure mode occurring. This evaluation relies on historical data, expert judgment, and statistical modeling. Among statistical techniques utilized, Bayesian networks can model the probability distributions over possible failure scenarios, enhancing the occurrence prediction accuracy.

3 Detection Quality

Detection (D) reflects the probability that a failure mode will be detected before it results in undesirable outcomes. A lower detection score signifies a greater likelihood of identifying the failure mode promptly, thus allowing for corrective action. Techniques such as Kalman filters and anomaly detection algorithms are employed to enhance detection capabilities in real-time monitoring systems.

Data-Driven Approaches in FMEA

Incorporating data-driven approaches into FMEA involves utilizing machine learning algorithms and data analytics to improve the accuracy of failure predictions and the optimization of system reliability. By leveraging large datasets collected from propulsion system operations, data-driven FMEA methodologies refine parameter estimates and update failure probability distributions.

1 Machine Learning in FMEA

Machine learning models, such as neural networks and support vector machines, are employed to detect patterns and correlations in failure data that may not be immediately discernible through traditional analysis methods. These models predict failure occurrences by learning from historical data and can improve the timeliness and precision of reliability assessments.

2 Probabilistic Risk Assessment

Probabilistic risk assessment (PRA) provides a quantitative basis for FMEA by employing probability theory and statistical methods to evaluate safety and reliability. The integration of PRA with FMEA enhances the understanding of complex component interdependencies and facilitates more robust risk mitigation strategies.

Let $P(F)$ represent the probability of failure, which can be decomposed using the law of total probability to incorporate multiple factors or conditions denoted as C_i:

$$P(F) = \sum_i P(F \mid C_i) \cdot P(C_i)$$

where $P(F \mid C_i)$ is the conditional probability of failure given condition C_i.

Implementation in Aerospace Propulsion Systems

Applying FMEA in aerospace propulsion systems involves meticulous documentation and analysis of each component's potential failure modes. This analysis is integral to the development and validation phases of system design, ensuring that risk mitigation strategies are incorporated from the outset. Continuous feedback loops that integrate operational data help evolve FMEA practices to adapt to changing failure environments and to maintain system reliability standards throughout the lifecycle of the system.

Python Code Snippet

Below is a Python code snippet that encompasses the core computational elements of Failure Mode and Effect Analysis (FMEA) for propulsion systems, including calculations for Risk Priority Numbers (RPN), Bayesian networks for occurrence likelihood, detection improvements with Kalman filters, and probabilistic risk assessment.

```python
import numpy as np
from scipy.stats import norm
from sklearn.naive_bayes import GaussianNB

def calculate_RPN(severity, occurrence, detection):
    '''
    Calculate the Risk Priority Number for a failure mode.
    :param severity: Severity score.
    :param occurrence: Occurrence score.
    :param detection: Detection score.
    :return: Risk Priority Number.
    '''
    return severity * occurrence * detection

def predict_failure_occurrence(data, labels):
    '''
    Predict failure occurrence using a Gaussian Naive Bayes model.
    :param data: Feature data of potential failure modes.
    :param labels: Known outcomes for training.
    :return: Trained model and prediction probabilities for input
    ↪  data.
    '''
    model = GaussianNB()
    model.fit(data, labels)
    predictions = model.predict(data)
    prediction_proba = model.predict_proba(data)
    return model, predictions, prediction_proba

def kalman_detection_update(measurement, estimate, uncertainty):
    '''
    Update detection quality using a Kalman filter algorithm.
    :param measurement: Current measurement data.
    :param estimate: Previous failure estimate.
    :param uncertainty: Uncertainty in the estimate.
    :return: Updated estimate and uncertainty.
    '''
    kalman_gain = uncertainty / (uncertainty + np.var(measurement))
    new_estimate = estimate + kalman_gain * (measurement - estimate)
    new_uncertainty = (1 - kalman_gain) * uncertainty
    return new_estimate, new_uncertainty

def probabilistic_risk_assessment(failure_conditions,
↪  probabilities):
    '''
    Calculate the probability of failure using probabilistic risk
    ↪  assessment.
    :param failure_conditions: Array of conditions.
    :param probabilities: Associated probabilities for each
    ↪  condition.
```

```
:return: Total probability of failure.
'''
    total_probability = np.sum([p * cond for p, cond in
    ↪ zip(probabilities, failure_conditions)])
    return total_probability

# Example implementation
severity_scores = [9, 8, 7]   # Sample severity scores for failure
↪ modes
occurrence_scores = [3, 4, 2]   # Sample occurrence scores
detection_scores = [2, 1, 1]   # Sample detection scores

rpn_values = [calculate_RPN(s, o, d) for s, o, d in
↪ zip(severity_scores, occurrence_scores, detection_scores)]

# Sample data and labels for the Gaussian Naive Bayes model
data = np.array([[0.1, 0.2], [0.2, 0.3], [0.3, 0.4]])
labels = np.array([1, 0, 1])
model, predictions, prediction_proba =
↪ predict_failure_occurrence(data, labels)

# Kalman filter applied to detection estimation
measurement = np.array([0.5, 0.6])
estimate = np.array([0.4, 0.5])
uncertainty = np.array([0.1, 0.1])
new_estimate, new_uncertainty = kalman_detection_update(measurement,
↪ estimate, uncertainty)

# Probabilistic risk assessment example
failure_conditions = [1, 0, 1]   # Hypothetical conditions
probabilities = [0.8, 0.5, 0.3]
total_failure_probability =
↪ probabilistic_risk_assessment(failure_conditions, probabilities)

print("RPN Values:", rpn_values)
print("Prediction Probabilities:", prediction_proba)
print("Updated Estimate:", new_estimate)
print("Total Failure Probability:", total_failure_probability)
```

This code defines several key functions necessary for the application of FMEA and risk assessment in propulsion systems:

- calculate_RPN function computes the Risk Priority Number based on severity, occurrence, and detection scores.

- predict_failure_occurrence uses a Gaussian Naive Bayes model to assess failure occurrence probabilities from data.

- kalman_detection_update applies a Kalman filter to improve detection quality and update estimates.

- `probabilistic_risk_assessment` evaluates the probability of failure using a structured probabilistic framework.

The final block of code provides examples of computing these elements using hypothetical data inputs.

Chapter 52

Gas Dynamics in Variable Cross-Sections

Fundamental Principles of Gas Dynamics

Gas dynamics involves the study of gas flow behavior, particularly when subject to changing duct geometries. The continuity equation, momentum equation, and energy equation form the core set of governing equations. For steady, one-dimensional flow through a duct of variable cross-section, the conservation of mass can be expressed as:

$$\dot{m} = \rho A u = \text{constant}$$

where $\dot{m}$ is the mass flow rate, ρ is the fluid density, A is the cross-sectional area, and u is the flow velocity.

Application of Bernoulli's Equation

Bernoulli's equation provides a relationship between pressure, velocity, and elevation in an inviscid, incompressible flow. For a streamline in a duct with no significant change in height, the equation is simplified to:

$$P + \frac{1}{2}\rho u^2 = \text{constant}$$

where P represents the static pressure. In regions where the duct area changes, the velocity of the gas must adjust to conserve energy, suggesting an inverse relationship between velocity and area for subsonic flow.

Isentropic Flow Relations

In many aerospace applications, the flow can be considered isentropic, implying adiabatic and reversible processes. The fundamental relations connecting pressure, density, and temperature in such flows are given by:

$$\frac{P_2}{P_1} = \left(\frac{\rho_2}{\rho_1}\right)^\gamma = \left(\frac{T_2}{T_1}\right)^{\frac{\gamma}{\gamma-1}}$$

where γ is the specific heat ratio. The isentropic relations are essential for predicting the behavior of compressible flows in variable cross-section ducts like convergent-divergent nozzles.

1 Mach Number Variation

The Mach number defines the compressibility of the flow within varying cross-sectional geometries. The differential form of the area-Mach number relation for an isentropic flow is:

$$\frac{dA}{A} = (M^2 - 1)\frac{du}{u}$$

indicating that for subsonic flow ($M < 1$), an increase in cross-sectional area reduces the velocity, whereas in supersonic flow ($M > 1$), an increase in area results in acceleration.

Shock Waves and Expansion Fans

In ducts with variable geometry, the flow may become supersonic, necessitating the consideration of shock waves and expansion fans. For normal shocks, the Rankine-Hugoniot relations relate conditions across the shock front:

$$\frac{\rho_2}{\rho_1} = \frac{(\gamma+1)M_1^2}{2 + (\gamma-1)M_1^2}$$

$$\frac{P_2}{P_1} = 1 + \frac{2\gamma}{\gamma+1}(M_1^2 - 1)$$

where subscripts 1 and 2 denote upstream and downstream conditions, respectively.

Nozzle Design and Optimization

The design and optimization of nozzles involve carefully shaping the duct cross-section to manage the transition from subsonic to supersonic flow regimes. For optimal expansion, the exit Mach number M_e should meet mission-specific requirements while minimizing losses. The thrust produced by the nozzle depends critically on its ability to efficiently convert pressure and thermal energy into kinetic energy.

For a perfect gas, the thrust F can be expressed as:

$$F = \dot{m} u_e + (P_e - P_a) A_e$$

where u_e is the exit velocity, P_e the exit pressure, P_a the ambient pressure, and A_e the exit area. Achieving appropriate flow expansion necessitates precise tuning of the duct's convergent and divergent sections, ensuring the nozzle operates efficiently across its design range.

1 Design Considerations

Key nozzle design considerations include the choice between fixed and variable geometries and the implications of flow separation or boundary layer effects. The Reynolds number informs about laminar and turbulent flow regimes, affecting boundary layer thickness and separation tendencies, crucial factors in nozzle efficiency.

The Prandtl-Meyer expansion provides a framework for calculating flow turning in supersonic expansion fans, imparting further insight into shock-expansion interactions:

$$\nu(M) = \sqrt{\frac{\gamma+1}{\gamma-1}} \arctan\left(\sqrt{\frac{\gamma-1}{\gamma+1}(M^2-1)}\right) - \arctan\left(\sqrt{M^2-1}\right)$$

where ν is the Prandtl-Meyer function. This equation is instrumental in defining the turning angle as a function of Mach number during expansion processes.

Python Code Snippet

Below is a Python code snippet that encompasses the core computational elements discussed in gas dynamics involving variable cross-sections, including the application of fundamental gas dynamics principles, Bernoulli's equation, isentropic flow relations, Mach number variations, and the calculation of thrust.

```
import numpy as np

def mass_flow_rate(rho, A, u):
    '''
    Calculate the mass flow rate of gas.
    :param rho: The fluid density.
    :param A: The cross-sectional area.
    :param u: The flow velocity.
    :return: Mass flow rate.
    '''
    return rho * A * u

def bernoulli_pressure(P1, rho, u1, u2):
    '''
    Apply Bernoulli's equation to calculate pressure changes.
    :param P1: Initial static pressure.
    :param rho: Fluid density.
    :param u1: Initial velocity.
    :param u2: Final velocity.
    :return: Final static pressure.
    '''
    return P1 + 0.5 * rho * (u1**2 - u2**2)

def isentropic_relations(P1, T1, M1, M2, gamma):
    '''
    Calculate isentropic flow relations between two states.
    :param P1: Initial pressure.
    :param T1: Initial temperature.
    :param M1: Initial Mach number.
    :param M2: Final Mach number.
    :param gamma: Specific heat ratio.
    :return: Tuple of final pressure and temperature.
    '''
    T2_T1 = (1 + 0.5 * (gamma - 1) * M1**2) / (1 + 0.5 * (gamma - 1)
     ↪ * M2**2)
    P2_P1 = T2_T1**(gamma / (gamma - 1))
    return P1 * P2_P1, T1 * T2_T1

def shock_wave_relations(M1, gamma):
    '''
    Calculate the Rankine-Hugoniot relations for a normal shock
     ↪ wave.
    :param M1: Upstream Mach number.
```

```
    :param gamma: Specific heat ratio.
    :returns: Tuple of density ratio and pressure ratio.
    '''
    rho_ratio = ((gamma + 1) * M1**2) / (2 + (gamma - 1) * M1**2)
    P_ratio = 1 + (2 * gamma * (M1**2 - 1)) / (gamma + 1)
    return rho_ratio, P_ratio

def thrust(m_dot, u_e, P_e, P_a, A_e):
    '''
    Calculate the thrust produced by a nozzle.
    :param m_dot: Mass flow rate.
    :param u_e: Exit velocity.
    :param P_e: Exit pressure.
    :param P_a: Ambient pressure.
    :param A_e: Exit area.
    :return: Thrust.
    '''
    return m_dot * u_e + (P_e - P_a) * A_e

# Example data for demonstration
rho = 1.225
A = 0.1
u1 = 200
u2 = 250
P1 = 101325
gamma = 1.4
M1 = 0.8
M2 = 1.2
m_dot = mass_flow_rate(rho, A, u1)
P2 = bernoulli_pressure(P1, rho, u1, u2)
P2_isentropic, T2_isentropic = isentropic_relations(P1, 300, M1, M2,
    ↪ gamma)
rho_ratio, P_ratio = shock_wave_relations(M1, gamma)

# Demonstrate thrust calculation
u_e = 300
P_e = 100000
P_a = 101325
A_e = 0.05
F = thrust(m_dot, u_e, P_e, P_a, A_e)

print("Mass Flow Rate:", m_dot)
print("Bernoulli Final Pressure:", P2)
print("Isentropic Final Pressure:", P2_isentropic)
print("Isentropic Final Temperature:", T2_isentropic)
print("Shock Wave Density Ratio:", rho_ratio)
print("Shock Wave Pressure Ratio:", P_ratio)
print("Thrust:", F)
```

This code defines several key functions necessary for evaluating gas dynamics in variable cross-section ducts:

- `mass_flow_rate` function computes the mass flow rate given density, area, and velocity.

- `bernoulli_pressure` calculates changes in pressure using Bernoulli's equation for inviscid flow.

- `isentropic_relations` determines final state variables using isentropic flow relations.

- `shock_wave_relations` applies Rankine-Hugoniot relations to find conditions across a shock.

- `thrust` calculates the thrust from a nozzle based on exit conditions.

The provided code block demonstrates these calculations with example values to illustrate their use in practical scenarios.

Chapter 53

Boundary Conditions in Computational Fluid Dynamics (CFD)

Introduction to Boundary Conditions in CFD

Boundary conditions in Computational Fluid Dynamics (CFD) are essential to the accurate and reliable prediction of aerodynamic behaviors. These conditions represent the physical constraints applied at the domain boundaries, determining the solution along the edges as well as influencing the entire flow field. Each type of boundary condition has specific formulations, ensuring that the CFD simulations conform to physical laws and address specific aerodynamic challenges in various configurations.

Types of Boundary Conditions

1 Dirichlet Boundary Condition

In CFD, the Dirichlet boundary condition, often referred to as a "fixed" or "essential" boundary condition, prescribes the value of a variable directly. For example, at a solid wall, the velocity components are set to zero to enforce the no-slip condition:

$$\mathbf{u}\big|_{\text{wall}} = 0 \qquad (53.1)$$

Additionally, temperature boundary conditions might be specified under the Dirichlet type when the temperature at the boundary is held constant, defined as:

$$T\big|_{\text{boundary}} = T_{fixed} \qquad (53.2)$$

2 Neumann Boundary Condition

The Neumann boundary condition is applied to the derivative of a variable, which often represents fluxes across boundaries. For instance, specifying a heat flux across a boundary involves setting:

$$\frac{\partial T}{\partial n}\bigg|_{\text{boundary}} = q \qquad (53.3)$$

where q represents the prescribed heat flux and $\frac{\partial T}{\partial n}$ is the normal derivative of temperature at the boundary. This type is crucial for inflow and outflow boundaries where pressure-gradient-driven flows are expected.

3 Robin Boundary Condition

The Robin boundary condition, a combination of Dirichlet and Neumann conditions, is less common but useful where coupled value and gradient predictions are necessary. It is represented as:

$$aT + b\frac{\partial T}{\partial n} = g \qquad (53.4)$$

This equation indicates a balance between the value of the variable and its gradient, controlled by the constants a and b, providing flexibility in simulating complex thermal and fluid interactions.

4 Periodic Boundary Condition

Periodic boundary conditions are deployed when the flow is expected to repeat at the boundaries, effectively reducing computational cost by simulating smaller domains. The equation for periodicity is:

$$\phi\big|_{\text{boundary 1}} = \phi\big|_{\text{boundary 2}} \qquad (53.5)$$

where ϕ represents any flow variable. This condition is pivotal in simulating flows in repeated geometries such as turbine blades and channels.

Implementation in CFD Solvers

Realization of boundary conditions within CFD simulations depends on the solver strategies and discretization approaches used. Most solvers convert continuous boundary conditions into discrete forms suitable for computational domains. Careful encoding of these conditions is necessary to maintain the fidelity of numerical predictions.

For a given velocity vector $\mathbf{u} = (u, v, w)$, a typical numerical discretization at a boundary node could be implemented using schemes like:

$$\frac{u_{i+1} - u_i}{\Delta x} = 0 \tag{53.6}$$

This simplistic form ensures that spatial derivatives and thus gradients are correctly represented within the discrete momentum equations.

Challenges and Considerations

Correct application of boundary conditions often involves reevaluating the physical context and potential perturbations. Common challenges include:

1 Alignment to Physical Laws

Ensuring that boundary conditions adhere to conservation laws such as mass, momentum, and energy conservation is significant. Failure in maintaining these can lead to non-physical results within the flow field.

2 Incompatible Domain Configurations

Complex geometries may necessitate hybrid boundary conditions or locally refined mesh zones where a particular type of boundary condition takes precedence based on local Reynolds numbers or Mach numbers.

3 Numerical Stability and Convergence

Boundary conditions often influence stability and convergence of CFD solutions. Unstable conditions may stem from improperly applied boundaries leading to non-convergent solutions, necessitating iterative refinement of the boundary parameters and grid resolution.

In conclusion, the selection and implementation of boundary conditions in CFD directly influence the accuracy and predictive capability of aerodynamic analyses, with particular attention paid to their alignment with physical phenomena, numerical fidelity, and domain-specific requirements.

Python Code Snippet

Below is a Python code snippet that encompasses key computational elements related to boundary conditions in CFD, including defining and implementing different types of boundary conditions in numerical simulations.

```
import numpy as np

def dirichlet_boundary_condition(velocity_field, boundary_nodes,
 ↪  value=0.0):
    '''
    Apply Dirichlet boundary condition by setting velocity to a
    ↪  fixed value.
    :param velocity_field: NumPy array representing the velocity
    ↪  field.
    :param boundary_nodes: Nodes where boundary condition is
    ↪  applied.
    :param value: Fixed value to set (default is no-slip condition,
    ↪  0).
    :return: Modified velocity field with applied boundary
    ↪  condition.
    '''
    for node in boundary_nodes:
        velocity_field[node] = value
    return velocity_field

def neumann_boundary_condition(temperature_field, boundary_nodes,
 ↪  flux):
    '''
    Apply Neumann boundary condition by specifying the derivative of
    ↪  temperature.
    :param temperature_field: NumPy array representing temperature
    ↪  field.
```

```
    :param boundary_nodes: Nodes where boundary condition is
    ↪ applied.
    :param flux: Prescribed heat flux.
    :return: Modified temperature field with applied boundary
    ↪ condition.
    '''
    for node in boundary_nodes:
        temperature_field[node] = temperature_field[node-1] + flux
    return temperature_field

def robin_boundary_condition(field, boundary_nodes, a, b, g):
    '''
    Apply Robin boundary condition for combining value and gradient.
    :param field: NumPy array of the field (e.g., temperature or
    ↪ velocity).
    :param boundary_nodes: Nodes where boundary condition is
    ↪ applied.
    :param a, b, g: Constants for Robin condition.
    :return: Modified field with applied boundary condition.
    '''
    for node in boundary_nodes:
        field[node] = (g - b * (field[node+1] - field[node])) / a
    return field

def periodic_boundary_condition(field, boundary_node1,
↪ boundary_node2):
    '''
    Apply periodic boundary condition by matching values at two
    ↪ boundaries.
    :param field: NumPy array of the field (e.g., velocity or
    ↪ pressure).
    :param boundary_node1, boundary_node2: Boundary nodes to match.
    :return: Modified field with applied boundary condition.
    '''
    field[boundary_node1] = field[boundary_node2]
    return field

# Example setup
velocity = np.zeros(10)   # Example velocity field
temperature = np.linspace(300, 350, 10)   # Example temperature field

# Nodes boundaries
no_slip_nodes = [0, -1]   # Dirichlet, no-slip condition at start and
↪ end
heat_flux_nodes = [0]   # Neumann, heat flux at start

# Apply boundary conditions
velocity = dirichlet_boundary_condition(velocity, no_slip_nodes)
temperature = neumann_boundary_condition(temperature,
↪ heat_flux_nodes, flux=5)

# Outputs
print("Velocity Field With Dirichlet Condition:", velocity)
```

```
print("Temperature Field With Neumann Condition:", temperature)
```

This code defines several key functions for implementing boundary conditions in CFD simulations:

- `dirichlet_boundary_condition` function sets a fixed value for the velocity field at specified nodes to impose a Dirichlet condition (e.g., no-slip at the wall).

- `neumann_boundary_condition` specifies a flux across a boundary, modifying the temperature field based on the gradient requirement.

- `robin_boundary_condition` combines Dirichlet and Neumann conditions for cases requiring coupled value and gradient constraints.

- `periodic_boundary_condition` ensures periodicity by matching field values at two boundaries, reducing computational domain size for repetitive geometries.

The final block of code illustrates how to apply these boundary conditions to a sample velocity and temperature field, showcasing their effect on computational domains.

Chapter 54

Vortex Dynamics in Rocket Exhaust

Fundamentals of Vortex Dynamics

The dynamics of vortices within rocket exhaust flows present a complex interaction of fluid motion, critical to understanding the disturbances these vortices may introduce. The underlying principles are rooted in the conservation laws of mass, momentum, and energy, as well as the vorticity equation. Vorticity, $\boldsymbol{\omega} = \nabla \times \mathbf{u}$, serves as a pivotal concept, denoting the local spinning motion of a fluid element, where $\mathbf{u}$ is the velocity field.

The Navier-Stokes equations govern the fluid flow, and for an incompressible, viscous flow, they are expressed as:

$$\frac{\partial \mathbf{u}}{\partial t} + (\mathbf{u} \cdot \nabla)\mathbf{u} = -\frac{1}{\rho}\nabla p + \nu \nabla^2 \mathbf{u}$$

where ρ represents fluid density, p is the pressure, and ν is the kinematic viscosity. These equations form the basis for understanding vortex formation and development in rocket exhausts.

Vortex Formation in Rocket Exhaust

The formation of vortices in rocket exhaust primarily stems from shear layer instabilities and pressure gradients. Shear layers occur at the edges of the exhaust jet, where velocity differences create a Kelvin-Helmholtz instability, leading to the development of coherent vortex structures. The growth of such instabilities can be described through the vorticity transport equation derived from the Navier-Stokes system:

$$\frac{D\boldsymbol{\omega}}{Dt} = (\boldsymbol{\omega} \cdot \nabla)\mathbf{u} - \boldsymbol{\omega}(\nabla \cdot \mathbf{u}) + \nu \nabla^2 \boldsymbol{\omega}$$

Mathematical Modeling of Vortex Dynamics

The mathematical modeling of vortex dynamics involves capturing the influence of geometrical and boundary condition variations on the flow field. Understanding the vortex shedding frequency and amplitude is crucial, often described using the Strouhal number, St, defined as:

$$\text{St} = \frac{fL}{U}$$

where f is the shedding frequency, L is a characteristic length, and U is the flow velocity. This nondimensional parameter provides insights into the flow patterns and scaling relationships characteristic of vortex formation.

Techniques for Mitigating Vortex-Induced Disturbances

Strategies to mitigate vortex-induced disturbances in rocket exhaust range from geometrical modifications to active flow control mechanisms. Suppression techniques focus on altering the conducive conditions for vortex formation. Geometric perturbations, such as chevrons or scalloped geometries, introduce controlled disturbances to the shear layers, breaking down vortex coherence. Such modifications can be evaluated through specific cost functions, J, reflecting total disturbance energy:

$$J = \int_V \left(\frac{1}{2} \rho |\mathbf{u}|^2 + p \right) dV$$

The aim is to minimize J through appropriate geometrical and operational settings, reducing the effective amplitude and impact of vortex shedding.

Numerical and Computational Approaches

Numerical simulations play a central role in analyzing and predicting the behavior of vortices in exhaust flows, typically employing Computational Fluid Dynamics (CFD) and Large Eddy Simulations (LES) for high-fidelity modeling. These methods resolve large, energy-containing eddies while modeling smaller turbulence scales. The filtering operation for LES is governed by the expression:

$$\widetilde{\mathbf{u}} = \int G(\mathbf{x} - \mathbf{x}')\mathbf{u}(\mathbf{x}')d\mathbf{x}'$$

where $\widetilde{\mathbf{u}}$ is the filtered velocity field, and G is the filter function. Such advanced techniques aid in the in-depth exploration and control of vortical structures, capturing the intricate details of exhaust dynamics critical to aerospace applications.

Python Code Snippet

Below is a Python code snippet that encompasses the core computational elements related to vortex dynamics, including vortex shedding frequency calculation, energy measurement of disturbances, and high-fidelity simulation setup using Computational Fluid Dynamics (CFD) techniques.

```python
import numpy as np
from scipy.integrate import quad

def vorticity(velocity_field, grid_spacing):
    '''
    Calculate the vorticity of a given velocity field.
    :param velocity_field: 2D array representing fluid velocities.
    :param grid_spacing: Distance between grid points.
    :return: Vorticity field as 2D array.
    '''
    du_dy, du_dx = np.gradient(velocity_field, grid_spacing)
    return np.subtract(du_dx, du_dy)

def strouhal_number(shedding_frequency, characteristic_length,
    flow_velocity):
    '''
    Calculate the Strouhal number.
    :param shedding_frequency: Frequency of vortex shedding.
    :param characteristic_length: Characteristic length of the
        object.
    :param flow_velocity: Velocity of the fluid flow.
```

```
    :return: Strouhal number.
    '''
    return (shedding_frequency * characteristic_length) /
    ↪    flow_velocity

def disturbance_energy_density(velocity_field, pressure_field,
↪  density):
    '''
    Calculate the disturbance energy density.
    :param velocity_field: 2D array for fluid velocity.
    :param pressure_field: 2D array for fluid pressure.
    :param density: Fluid density.
    :return: Disturbance energy density.
    '''
    kinetic_energy_density = 0.5 * density *
    ↪    np.sum(np.power(velocity_field, 2))
    potential_energy_density = np.sum(pressure_field)
    return kinetic_energy_density + potential_energy_density

def LES_filter(velocity_field, filter_function, grid_points):
    '''
    Apply a filter to the velocity field for Large Eddy Simulation.
    :param velocity_field: 2D array representing velocity field.
    :param filter_function: Function defining the filter.
    :param grid_points: Array of grid positions.
    :return: Filtered velocity field.
    '''
    filtered_velocity = np.zeros_like(velocity_field)
    for i, x in enumerate(grid_points):
        filtered_values = [filter_function(x - xp) * up for xp, up
        ↪    in zip(grid_points, velocity_field)]
        filtered_velocity[i] = np.sum(filtered_values)
    return filtered_velocity

# Example usage with dummy data
velocity_field = np.array([[1, 2, 3], [4, 5, 6], [7, 8, 9]])
pressure_field = np.array([[10, 10, 10], [20, 20, 20], [30, 30,
↪  30]])
grid_spacing = 1.0

vort_field = vorticity(velocity_field, grid_spacing)
strouhal = strouhal_number(5, 0.1, 20)
energy_density = disturbance_energy_density(velocity_field,
↪  pressure_field, 1.225)

def gauss_filter(x):
    return np.exp(-0.5*x**2)

filtered_velocity = LES_filter(velocity_field, gauss_filter,
↪  np.linspace(-1, 1, len(velocity_field)))

print("Vorticity Field:", vort_field)
print("Strouhal Number:", strouhal)
```

```
print("Disturbance Energy Density:", energy_density)
print("Filtered Velocity Field:", filtered_velocity)
```

This code defines several key functions necessary for understanding and simulating the dynamics of vortices in rocket exhaust systems:

- `vorticity` function computes the vorticity field from a given velocity field using finite differences.

- `strouhal_number` calculates the Strouhal number, indicating vortex shedding characteristics.

- `disturbance_energy_density` evaluates the disturbance energy density by combining kinetic and potential energy densities.

- `LES_filter` demonstrates appliance of a filtering function, essential for Large Eddy Simulations, to resolve larger flow structures.

The final block of code provides examples of computing these elements using hypothetical data for demonstration purposes.

Chapter 55

Dynamic Stability Margin Calculations

Dynamic Stability Fundamentals

In the realm of aerospace engineering, the dynamic stability of a rocket is a critical facet that ensures the vehicle's ability to maintain its course and perform necessary corrections without deviating undesirably from its intended path. The concept of dynamic stability hinges on the understanding of the vehicle's response to perturbations during flight and the subsequent damping characteristics that either restore or exacerbate these disturbances.

Dynamic stability is typically assessed through the linearized equations of motion that describe a rocket's response to small perturbations. The essential representation is given by a set of linear differential equations:

$$\dot{\mathbf{x}} = \mathbf{A}\mathbf{x} + \mathbf{B}\mathbf{u}$$

where $\mathbf{x}$ represents the state vector encompassing variables such as angular velocities and angular displacements, $\mathbf{u}$ denotes the control input vector, $\mathbf{A}$ is the state matrix, and $\mathbf{B}$ is the input matrix. The eigenvalues of the state matrix $\mathbf{A}$ dictate the stability characteristics, where the real parts must necessarily be negative for the system to be dynamically stable.

Defining Stability Margins

The stability margins in dynamic systems quantify the degree to which the system can tolerate parameter variations before losing stability. The gain margin and phase margin are fundamental in characterizing dynamic stability within the frequency domain, typically evaluated using Bode plots or Nyquist criteria. These margins are pivotal in ensuring robustness against system uncertainties and environmental disturbances.

The gain margin G_m can be defined as:

$$G_m = \frac{1}{|G(j\omega_{180})|}$$

where $G(j\omega_{180})$ is the system's open-loop transfer function evaluated at the phase crossover frequency, ω_{180}.

The phase margin ϕ_m is calculated by:

$$\phi_m = 180° + \arg(G(j\omega_c))$$

where ω_c is the gain crossover frequency, the frequency at which the magnitude of the open-loop transfer function equals unity.

Computing Dynamic Stability Margins

The dynamic stability margins for rockets during flight are assessed over a spectrum of operating conditions, including various altitudes and velocities, thereby ensuring adaptability and control. The application of the `rootlocus` technique is instrumental in visualizing pole movements as system parameters are varied. The characteristic equation is expressed as:

$$\det(s\mathbf{I} - \mathbf{A} + k\mathbf{B}) = 0$$

where s is the complex frequency variable, and k is the feedback gain.

Control design inherently relies on ensuring that these poles are appropriately placed within the left half of the complex plane for guaranteed stability. The pole placement or `pole` must be done such that:

$$\text{Re}(s_i) < 0, \quad \forall s_i \in \sigma(\mathbf{A} - k\mathbf{B})$$

Techniques for Margin Enhancement

Marginal enhancement techniques involve structural and control modifications adapted to minimize the risk of instability. These include adjustments to the center of gravity, the distribution of mass, and the use of gyroscopic controls to improve dynamic response. Feedback mechanisms, employing techniques like `PID` control or `H-infinity` optimization, aim to refine the dynamic response by altering loop gains and phase responses.

Dynamic inversion is another sophisticated method, utilizing models to counteract dynamic variations pre-emptively, rendering the nonlinearities of the system into a linearized equivalent. The inversion technique manipulates input variables such that:

$$\mathbf{u} = -(\mathbf{B}^T\mathbf{B})^{-1}\mathbf{B}^T\mathbf{f}(\mathbf{x})$$

thereby achieving desired outputs irrespective of internal and external perturbations.

Example Application

Consider a launch scenario where stability margins need evaluation under perturbative conditions caused by environmental turbulences. The state-space representation must accommodate the rocket's inherent nonlinearities via linearization about a nominal operating point. Variations in thrust vector control, T_c, are managed to enhance stability margins, applying classical feedback control laws:

$$T_c = -K(\mathbf{x} - \mathbf{x}_{desired})$$

Optimization methods converge on a gain matrix K which ensures poles' stability, empirically verified through simulation or on-the-fly adjustments during early flight trials.

Python Code Snippet

Below is a Python code snippet that encompasses the core computational elements for assessing dynamic stability margins of a rocket. This includes the calculation of eigenvalues for stability, gain, and phase margins determination, and pole placement using control system techniques.

```python
import numpy as np
import control

def calculate_eigenvalues(A):
    '''
    Calculate the eigenvalues of the state matrix A to assess
    stability.
    :param A: State matrix.
    :return: Eigenvalues of the matrix.
    '''
    eigenvalues = np.linalg.eigvals(A)
    return eigenvalues

def stability_margins(system):
    '''
    Calculate gain and phase margins for the given system.
    :param system: Control system transfer function.
    :return: Gain and phase margins.
    '''
    gm, pm, sm, wg, wp, ws = control.stability_margins(system)
    return gm, pm, wg, wp

def root_locus(system):
    '''
    Plot the root locus of the control system.
    :param system: Control system transfer function.
    :return: Root locus plot data.
    '''
    return control.root_locus(system, Plot=True)

def pole_placement(A, B, desired_poles):
    '''
    Calculate the feedback gain matrix for given desired poles.
    :param A: State matrix.
    :param B: Input matrix.
    :param desired_poles: Desired location for system poles.
    :return: Gain matrix K.
    '''
    K = control.place(A, B, desired_poles)
    return K

def dynamic_inversion(f, B, x):
    '''
    Apply dynamic inversion to compensate for non-linearities.
    :param f: Nonlinear dynamics function.
    :param B: Input matrix.
    :param x: Current state vector.
    :return: Compensated control input.
    '''
    B_inv_transpose = np.linalg.pinv(B.T @ B) @ B.T
    u = - B_inv_transpose @ f(x)
    return u
```

```python
# Define state-space representation
A = np.array([[0, 1], [-2, -3]])
B = np.array([[0], [1]])
C = np.array([[1, 0]])
D = np.array([[0]])

# Define the control system
sys = control.ss(A, B, C, D)

# Eigenvalues for dynamic stability check
eigenvalues = calculate_eigenvalues(A)
print("Eigenvalues of the system:", eigenvalues)

# Calculate stability margins
gm, pm, wg, wp = stability_margins(control.ss2tf(sys))
print(f"Gain Margin: {gm}, Phase Margin: {pm}\n")
print(f"Gain Crossover Frequency: {wg}, Phase Crossover Frequency:
↪ {wp}\n")

# Root locus for visualizing pole movement
root_locus(sys)

# Perform pole placement for desired stability properties
desired_poles = [-2, -4]   # desired stable pole locations
K = pole_placement(A, B, desired_poles)
print("Feedback gain matrix K for desired pole placement:", K)

# Example nonlinear function for dynamic inversion
def example_nonlinear_dynamics(x):
    return np.array([x[1], -x[0]**2 + 3*x[1]])

# Calculate control input using dynamic inversion
x = np.array([1.0, 0.5])   # Example state
control_input = dynamic_inversion(example_nonlinear_dynamics, B, x)
print("Control input from dynamic inversion:", control_input)
```

This code defines several key functions necessary for analyzing and enhancing the dynamic stability of rocket systems:

- calculate_eigenvalues function computes the eigenvalues of the state matrix **A** to assess if the system is stable.

- stability_margins calculates the gain and phase margins using Bode analysis, important for maintaining stability against uncertainties.

- root_locus provides a visual representation of how poles migrate with changes in system parameters, aiding in stability analysis.

- `pole_placement` determines the feedback gain matrix **K** to achieve desired pole locations for improved stability.

- `dynamic_inversion` calculates the control input needed to counteract system nonlinearities based on the dynamic inversion technique.

The final block of code demonstrates the application of these functions with an example aerospace scenario, showcasing eigenvalue calculation, stability margins, pole placement, and dynamic inversion.

Chapter 56

Advanced Finite Element Analysis (FEA)

Introduction to Finite Element Analysis

Finite Element Analysis (FEA) constitutes an indispensable tool in aerospace engineering, facilitating the evaluation of structural integrity under multifaceted loading conditions. In the aerospace sector, structures encounter complex aerodynamic loads, thermal stresses, and dynamic forces, all of which necessitate a robust analysis methodology. The FEA approach subdivides structures into interconnected elements, wherein the material properties and governing equations of each element contribute to an overarching system of algebraic equations.

The fundamental equation constituting FEA is predicated on the principle of minimum potential energy, expressed as:

$$\int_V \sigma\,\epsilon\,dV = \int_V \rho\,\mathbf{u}\cdot\mathbf{f}\,dV + \int_S \mathbf{u}\cdot\mathbf{t}\,dS$$

where σ is the stress tensor, ϵ the strain tensor, ρ the material density, $\mathbf{u}$ the displacement vector, $\mathbf{f}$ the volumetric force vector, and $\mathbf{t}$ the surface traction vector.

Formulation of Element Stiffness Matrix

The core component of FEA is the formation of the element stiffness matrix $\mathbf{K}_e$. The stiffness matrix represents how the displacements of nodes result in internal forces. For a linear elastic isotropic material, the stiffness matrix of an element is described as:

$$\mathbf{K}_e = \int_{V_e} \mathbf{B}^T \mathbf{D} \mathbf{B}\, dV_e$$

Here, $\mathbf{B}$ is the strain-displacement matrix that defines how nodal displacements translate into element strains, and $\mathbf{D}$ is the material property matrix, often referred to as the constitutive matrix for isotropic materials, typically filled with the elastic modulus E and Poisson's ratio ν.

Meshing Techniques in Aerospace Applications

Meshing forms the crux of an accurate FEA model, utilizing a variety of element shapes such as tetrahedra, hexahedra, and shell elements to discretize aerospace components. The resolution of the mesh significantly influences the precision of stress and deflection predictions.

In aerospace applications, regions with anticipated stress concentrations, such as fastener holes and sharp geometries, necessitate refined meshing. Adaptive meshing techniques, which iteratively refine certain areas based on preliminary stress analysis, can enhance accuracy without excessively increasing the computational load.

Non-linear FEA Challenges

Non-linear FEA addresses complex material and geometric behaviors introduced in aerospace structures. Material non-linearities include plastic deformations where the stress-strain relationship becomes non-linear beyond the yield point:

$$\sigma = E_1 \cdot \epsilon, \quad \text{for } \epsilon < \epsilon_y \quad \sigma = E_2 \cdot (\epsilon - \epsilon_y) + \sigma_y, \quad \text{for } \epsilon \geq \epsilon_y$$

with E_1 as the elastic modulus, E_2 as a modulus in the plastic region, ϵ_y denoting yield strain, and σ_y as yield stress.

Geometric non-linearities arise from large deformations affecting the original equilibrium configuration. Solving such complexities often requires iterative methods, such as the Newton-Raphson technique to update the stiffness matrix after each increment:

$$\mathbf{R}_i = \mathbf{F}_{\text{external}} - \mathbf{K}_i\,\mathbf{u}_i$$

Where $\mathbf{R}_i$ is the residual force vector, and iterations continue until $|\mathbf{R}_i| <$ tolerance.

Dynamic Analysis in FEA

In aerospace structures, dynamic analysis is crucial for understanding the response to time-dependent loads such as buffeting and flutter. Techniques like modal analysis and direct time integration are utilized to evaluate these effects.

Modal analysis decomposes the structure's response into a series of natural vibration modes, expressed through an eigenvalue problem:

$$\mathbf{K}\,\boldsymbol{\Phi} = \lambda \mathbf{M}\,\boldsymbol{\Phi}$$

where $\mathbf{K}$ and $\mathbf{M}$ are the global stiffness and mass matrices, $\boldsymbol{\Phi}$ is the eigenvector matrix, and λ are the eigenvalues corresponding to the natural frequencies.

Conversely, direct time integration methods like the Newmark-β method afford solutions for transient dynamic response:

$$\mathbf{M}\,\ddot{\mathbf{u}} + \mathbf{C}\,\dot{\mathbf{u}} + \mathbf{K}\,\mathbf{u} = \mathbf{f}(t)$$

where $\mathbf{C}$ is the damping matrix.

Applications of FEA in Aerospace Design

Advanced FEA techniques permit the evaluation of cutting-edge aerospace components, such as composite wings, rocket fuselage designs, and intricate turbine engines. By synthesizing various types of analyses, engineers can predict failure modes, optimize weight and material use, and ensure compliance with rigid certification standards.

Through numerous simulations, planners assess yield points, buckling tendencies, and stress propagation under repeated loadings, directly contributing to innovations in lightweight structural designs and increasing the reliability and safety margins of aerospace vehicles.

Python Code Snippet

Below is a Python code snippet that elaborates on the implementation of core finite element analysis methodologies, including establishing the element stiffness matrix, performing modal analysis, calculating non-linear deformations, and dynamic response assessments.

```
import numpy as np
from scipy.linalg import eigh
from scipy.integrate import solve_ivp

def element_stiffness_matrix(B, D, V_e):
    '''
    Calculate the stiffness matrix for an element.
    :param B: Strain-displacement matrix.
    :param D: Material property matrix.
    :param V_e: Element volume.
    :return: Element stiffness matrix.
    '''
    return B.T @ D @ B * V_e

def modal_analysis(K, M):
    '''
    Perform modal analysis to find natural frequencies.
    :param K: Global stiffness matrix.
    :param M: Global mass matrix.
    :return: Natural frequencies and mode shapes.
    '''
    eigenvalues, eigenvectors = eigh(K, M)
    natural_frequencies = np.sqrt(eigenvalues)
    return natural_frequencies, eigenvectors

def newmark_beta_method(M, C, K, f, u0, v0, t_span, beta=0.25,
    ↪ gamma=0.5):
    '''
    Solve dynamic response using Newmark-beta method.
    :param M: Mass matrix.
    :param C: Damping matrix.
    :param K: Stiffness matrix.
    :param f: External force vector as a function of time.
    :param u0: Initial displacement.
    :param v0: Initial velocity.
    :param t_span: Time span for solution.
    :param beta: Newmark method parameter.
    :param gamma: Newmark method parameter.
    :return: Displacement and velocity over time.
    '''
    dt = t_span[1] - t_span[0]
    n_steps = len(t_span)
```

```python
    u = np.zeros((n_steps, len(u0)))
    v = np.zeros((n_steps, len(v0)))
    a = np.zeros((n_steps, len(u0)))

    u[0], v[0] = u0, v0
    effective_stiffness = K + gamma / beta / dt * C + 1 / beta / 
     ↪  dt**2 * M

    a[0] = np.linalg.inv(M) @ (f(0) - C @ v[0] - K @ u[0])

    for i in range(1, n_steps):
        time = t_span[i]
        effective_force = f(time) + M @ (1 / beta / dt**2 * u[i-1]
                            + 1 / beta / dt * v[i-1]
                            + (1 / 2 / beta - 1) * a[i-1]) + \
                        C @ (gamma / beta / dt * u[i-1]
                            + (gamma / beta - 1) * v[i-1]
                            + dt * (gamma / 2 / beta - 1) *
     ↪                       a[i-1])

        u[i] = np.linalg.solve(effective_stiffness, effective_force)
        a[i] = 1 / beta / dt**2 * (u[i] - u[i-1]) - 1 / beta / dt *
     ↪   v[i-1] - (1 / 2 / beta - 1) * a[i-1]
        v[i] = v[i-1] + dt * ((1 - gamma) * a[i-1] + gamma * a[i])

    return u, v

# Function definitions for stress-strain relationship
def linear_stress_strain(E, strain):
    '''
    Calculate stress for linear relationship.
    :param E: Elastic modulus.
    :param strain: Strain.
    :return: Stress.
    '''
    return E * strain

def plastic_stress_strain(E1, E2, epsilon_y, sigma_y, strain):
    '''
    Calculate stress for plastic relationship.
    :param E1: Elastic modulus.
    :param E2: Modulus in plastic region.
    :param epsilon_y: Yield strain.
    :param sigma_y: Yield stress.
    :param strain: Strain.
    :return: Stress.
    '''
    if strain <= epsilon_y:
        return E1 * strain
    else:
        return E2 * (strain - epsilon_y) + sigma_y

# Sample parameters for testing
```

```python
B = np.array([[1, 0], [0, 1]])
D = np.array([[210e9, 0], [0, 210e9]])
V_e = 0.01
K_e = element_stiffness_matrix(B, D, V_e)

K = K_e
M = np.identity(2) * 1e3

# Modal Analysis
natural_frequencies, mode_shapes = modal_analysis(K, M)

# Newmark-beta Dynamic Analysis
u0 = np.zeros(2)    # Initial displacement
v0 = np.zeros(2)    # Initial velocity
t_span = np.linspace(0, 10, 100)
def f(t): return np.array([0, 1000 * np.sin(2 * np.pi * 1 * t)])

displacements, velocities = newmark_beta_method(M, np.zeros_like(K),
    ↪ K, f, u0, v0, t_span)

print("Element Stiffness Matrix:", K_e)
print("Natural Frequencies:", natural_frequencies)
print("Displacements:", displacements)
```

This code defines key computational functions required for finite element analysis in aerospace engineering:

- element_stiffness_matrix calculates the stiffness matrix for given material and mesh properties.

- modal_analysis determines the natural frequencies and mode shapes through solving an eigenvalue problem.

- newmark_beta_method performs a direct time integration for dynamic studies, solving motion equations for transient analyses.

- linear_stress_strain and plastic_stress_strain provide mathematical implementations for stress-strain relationships under linear and plastic regimes.

The sample code at the end demonstrates these operations with a simplified setup for clarity and educational intent.

Chapter 57

Micro Thrusters and Satellite Attitude Control

Fundamentals of Attitude Control using Micro Thrusters

Micro thrusters are crucial components for precise attitude control in small satellites and deep space probes. By providing finely tuned thrust increments, they enable orientation adjustments essential for mission-critical operations such as imaging, communication, and scientific data acquisition. Attitude control systems rely on principles of rotational dynamics, as encapsulated in Euler's equations, given by:

$$\mathbf{I} \cdot \frac{d\boldsymbol{\omega}}{dt} + \boldsymbol{\omega} \times (\mathbf{I} \cdot \boldsymbol{\omega}) = \mathbf{N}$$

where $\mathbf{I}$ denotes the moment of inertia tensor, $\boldsymbol{\omega}$ the angular velocity vector, and $\mathbf{N}$ the external torque vector. The role of micro thrusters is to provide this torque $\mathbf{N}$, thereby altering the satellite's angular momentum.

Micro Thruster Types and Their Efficiency

Different types of micro thrusters, such as cold gas, electric, and chemical, offer diverse trade-offs between power, efficiency, and thrust capabilities. A primary metric of efficiency is specific impulse (I_{sp}), defined as:

$$I_{sp} = \frac{T}{\dot{m} g_0}$$

where T is the thrust, $\dot{m}$ the propellant mass flow rate, and g_0 the standard gravitational acceleration. Electric thrusters, including ion and Hall effect thrusters, offer high I_{sp} at the cost of reduced thrust levels, making them suitable for sustained, long-duration missions.

Propellant Mass and Thruster Sizing

The equation governing the required propellant mass for a specific mission profile involves delta-v (Δv) budgeting, expressed using the Tsiolkovsky Rocket Equation:

$$\Delta v = I_{sp} \cdot g_0 \cdot \ln\left(\frac{m_0}{m_f}\right)$$

where m_0 and m_f are the initial and final mass of the satellite, respectively. Ensuring optimal thruster sizing requires solving for the propellant mass $m_0 - m_f$, such that mission objectives are met within mass and volume constraints.

Control Strategies in Attitude Modulation

Attitude control strategy implementation hinges on feedback control loops. These loops incorporate sensor data to adjust thruster outputs, maintaining the desired orientation. The control law is often designed using Linear Quadratic Regulator (LQR) techniques, optimizing the state feedback gain $\mathbf{K}$ to minimize the cost function:

$$J = \int (\mathbf{x}^T \mathbf{Q} \mathbf{x} + \mathbf{u}^T \mathbf{R} \mathbf{u}) \, dt$$

where **x** is the state vector, **u** the control input vector, and **Q** and **R** are weight matrices determining the trade-offs between state error and control effort.

Mathematical Modeling of Micro Thruster Dynamics

The dynamic model of micro thrusters is pivotal for predicting system responses. Using state-space representation, the thruster dynamics can be characterized as:

$$\dot{\mathbf{x}} = \mathbf{A}\mathbf{x} + \mathbf{B}\mathbf{u}$$

$$\mathbf{y} = \mathbf{C}\mathbf{x} + \mathbf{D}\mathbf{u}$$

where **x** is the state vector of the system, **u** the control input (thrust commands), **y** the output (e.g., achieved orientation), and **A**, **B**, **C**, **D** are matrices defining the system configuration and control influence.

Challenges and Optimization of Micro Thruster Systems

Implementing micro thrusters necessitates overcoming several challenges, including minimizing jitter and achieving low-noise operation critical for high-precision tasks. Solutions involve finely tuning control systems through strategies such as Genetic Algorithm optimization to refine control parameters:

$$\text{minimize } F(\mathbf{p}) = \sum_{i=1}^{n} f_i(\mathbf{p})$$

subject to constraints on thrust level, I_{sp}, and mass. Optimization seeks an efficient design to balance performance and resource usage.

These equations and techniques underscore the theoretical underpinnings necessary for the practical application of micro thrusters in satellite attitude control systems, enabling successful navigation and maneuvering in space missions.

Python Code Snippet

Below is a Python code snippet that captures the core computational elements of micro thruster dynamics and satellite attitude control, including the calculation of Euler's equations for rotational dynamics, specific impulse, propellant mass, attitude control strategies, and genetic algorithm optimization.

```python
import numpy as np
from scipy.integrate import odeint
from scipy.linalg import solve_continuous_are

def euler_equations(I, omega, N):
    '''
    Solves the Euler's equations for rotational dynamics.
    :param I: Moment of inertia tensor.
    :param omega: Angular velocity vector.
    :param N: External torque vector.
    :return: Derivative of angular velocity.
    '''
    return np.linalg.inv(I) @ (N - np.cross(omega, I @ omega))

def specific_impulse(thrust, mass_flow_rate, g0=9.81):
    '''
    Calculate specific impulse.
    :param thrust: Thrust generated by the thruster.
    :param mass_flow_rate: Propellant mass flow rate.
    :param g0: Standard gravitational acceleration.
    :return: Specific impulse.
    '''
    return thrust / (mass_flow_rate * g0)

def tsilkovsky_rocket_equation(m0, mf, I_sp, g0=9.81):
    '''
    Calculate delta-v using Tsiolkovsky Rocket Equation.
    :param m0: Initial mass.
    :param mf: Final mass.
    :param I_sp: Specific impulse.
    :param g0: Standard gravitational acceleration.
    :return: Delta-v.
    '''
    return I_sp * g0 * np.log(m0 / mf)

def calc_lqr_gain(A, B, Q, R):
    '''
    Calculate the optimal state feedback gains using LQR.
    :param A: State-space A matrix.
    :param B: State-space B matrix.
    :param Q: State weight matrix.
    :param R: Control weight matrix.
    :return: State feedback gain matrix K.
```

```python
    '''
    X = solve_continuous_are(A, B, Q, R)
    K = np.linalg.inv(R) @ (B.T @ X)
    return K

def genetic_algorithm_optimization(f, bounds, n_generations=100,
    n_population=50):
    '''
    Optimize a function using a simple genetic algorithm.
    :param f: Fitness function to minimize.
    :param bounds: Bounds for the variables.
    :param n_generations: Number of generations.
    :param n_population: Population size.
    :return: Optimized parameters.
    '''
    population = np.random.rand(n_population, len(bounds)) * \
                 (np.array(bounds)[:,1] - np.array(bounds)[:,0]) + \
                    np.array(bounds)[:,0]

    for _ in range(n_generations):
        # Evaluate each solution
        fitness = np.array([f(ind) for ind in population])
        idx = np.argsort(fitness)
        population = population[idx]  # Sort population by fitness

        # Crossover
        for i in range(1, n_population, 2):
            crossover_point = np.random.randint(1, len(bounds))
            population[i, :crossover_point], population[i+1,
                :crossover_point] = population[i+1,
                :crossover_point].copy(), population[i,
                :crossover_point].copy()

        # Mutation
        mutation_probability = 0.1
        for ind in population:
            if np.random.rand() < mutation_probability:
                mutation_dim = np.random.randint(len(bounds))
                ind[mutation_dim] = \
                    np.random.uniform(bounds[mutation_dim][0],
                    bounds[mutation_dim][1])

    return population[0]  # Return the best solution

def state_space_dynamics(x, t, A, B, u):
    '''
    Simulate state-space dynamics using a linear model.
    :param x: State vector.
    :param t: Time.
    :param A: State matrix.
    :param B: Input matrix.
    :param u: Control input.
    :return: Derivative of state vector.
```

```
    '''
    return A @ x + B @ u

# Example usage of the functions
I = np.diag([10, 10, 10])    # Example inertia tensor
omega = np.array([0.1, 0.2, 0.3])    # Initial angular velocity
N = np.array([0.01, 0.02, 0.03])    # Torque produced by thrusters

domega_dt = euler_equations(I, omega, N)

thrust = 0.5    # N
mass_flow_rate = 0.01    # kg/s
I_sp = specific_impulse(thrust, mass_flow_rate)

m0 = 10.0    # kg
mf = 9.5    # kg
delta_v = tsilkovsky_rocket_equation(m0, mf, I_sp)

# System matrices for state feedback control
A = np.array([[0, 1], [0, 0]])
B = np.array([[0], [1]])
Q = np.eye(2)
R = np.eye(1)
K = calc_lqr_gain(A, B, Q, R)

# Genetic algorithm optimization example
def fitness_func(params):
    # Dummy fitness function
    return np.sum(params**2)

bounds = [(-1, 1), (-1, 1), (-1, 1)]
optimal_params = genetic_algorithm_optimization(fitness_func,
↪    bounds)

# Simulate state-space system dynamics
x0 = np.array([0, 1])
t = np.linspace(0, 10, 100)
u = np.array([0.1])
x = odeint(state_space_dynamics, x0, t, args=(A, B, u))

print("Change in Angular Velocity:", domega_dt)
print("Specific Impulse:", I_sp)
print("Delta-v:", delta_v)
print("LQR Gain Matrix K:", K)
print("Optimal Parameters from GA:", optimal_params)
```

This Python code defines several key functions necessary for the simulation and optimization of micro thruster and satellite attitude control systems:

- `euler_equations` computes the rotational dynamics using

Euler's equations based on inertia, angular velocity, and applied torque.

- `specific_impulse` calculates the efficiency of a thruster system in terms of specific impulse.

- `tsilkovsky_rocket_equation` evaluates the change in velocity given specific impulse and mass properties.

- `calc_lqr_gain` computes the optimal feedback gain matrix for state-space systems using LQR.

- `genetic_algorithm_optimization` is a simple example of a genetic algorithm applied for function optimization, demonstrating evolutionary computation techniques.

- `state_space_dynamics` simulates the behavior of a given state-space model under specific conditions.

The example usage section of the code demonstrates how to compute these metrics using sample data and model parameters.

Chapter 58

Boundary Layer Control in Rockets

Introduction to Boundary Layer Theory

The performance and stability of rockets are significantly influenced by boundary layer dynamics. The boundary layer, a thin region where viscous forces dominate, forms on the surface of the rocket as it moves through a fluid. The laminar and turbulent characteristics of this layer impact the skin friction drag experienced by the vehicle. The boundary layer thickness, δ, is central to understanding these effects and is determined by Reynolds number, Re, such that:

$$\delta \approx \frac{5x}{\sqrt{\text{Re}}}$$

where x is the distance from the leading edge. Turbulent boundary layers are characterized by increased momentum and energy exchange, leading to higher skin friction but improved mixing and heat transfer.

Techniques for Boundary Layer Control

Various boundary layer control methods are employed to reduce skin friction drag, thereby enhancing rocket performance. These include passive methods, such as surface treatment and shaping, and active methods, such as boundary layer suction and blowing.

1 Surface Treatment and Shaping

Surface modification techniques aim to delay boundary layer transition or promote laminar flow over the surface of the rocket. Utilizing riblets or compliant coatings can effectively reduce drag by altering the near-wall turbulence structure. Mathematical models indicate that the drag reduction is proportional to the wall shear stress, τ_w, and can be estimated by:

$$\Delta \tau_w = -C_f (1 - \phi)$$

where C_f is the skin friction coefficient and ϕ represents the effectiveness of the surface treatment.

2 Boundary Layer Suction and Blowing

Boundary layer suction involves removing low-momentum fluid from the boundary layer via perforated surfaces, effectively delaying transition and reducing drag. The suction velocity, v_s, plays a crucial role and can be optimized through:

$$\text{Re}_s = \frac{\rho v_s L}{\mu} < \text{Re}_{\text{critical}}$$

where Re_s is the suction Reynolds number, L is a characteristic length, ρ the fluid density, and μ the dynamic viscosity.

Blowing techniques introduce high-momentum air into the boundary layer, energizing it to resist separation. The momentum coefficient, C_μ, quantifies the effectiveness of blowing:

$$C_\mu = \frac{\dot{m}_b V_b}{\frac{1}{2} \rho U_\infty^2 Ac}$$

where $\dot{m}_b$ is the mass flow rate of the blown air, V_b the blowing velocity, U_∞ the free stream velocity, and Ac the cross-sectional area affected by blowing.

Impact on Rocket Performance

Enhanced boundary layer control directly affects rocket efficiency by minimizing drag forces and reducing fuel consumption. The reduction in skin friction drag contributes to the overall drag coefficient, C_D, expressed as:

$$C_D = C_{D_0} + \Delta C_{D_f}$$

where C_{D_0} is the baseline drag coefficient and ΔC_{D_f} represents modifications due to boundary layer control.

In conclusion, mastery of these techniques can lead to significant advancements in the aerodynamic design and operational efficiency of rockets, enabling more effective missions through improved performance metrics.

Python Code Snippet

Below is a Python code snippet that encompasses the core computational elements related to boundary layer control in rockets, including calculations for boundary layer thickness, skin friction drag reduction, and techniques such as boundary layer suction and blowing.

```python
import numpy as np

def boundary_layer_thickness(Re, x):
    '''
    Calculate the boundary layer thickness.
    :param Re: Reynolds number.
    :param x: Distance from the leading edge.
    :return: Boundary layer thickness.
    '''
    return 5 * x / np.sqrt(Re)

def skin_friction_reduction(C_f, phi):
    '''
    Calculate the change in wall shear stress due to surface
    ↪ treatment.
    :param C_f: Skin friction coefficient.
    :param phi: Effectiveness of the surface treatment.
    :return: Change in wall shear stress.
    '''
    return -C_f * (1 - phi)

def suction_velocity(Re_critical, rho, L, mu):
    '''
    Calculate the suction velocity for boundary layer control.
    :param Re_critical: Critical Reynolds number for transition.
    :param rho: Fluid density.
    :param L: Characteristic length.
    :param mu: Dynamic viscosity.
    :return: Suction velocity.
    '''
```

```python
    return (Re_critical * mu) / (rho * L)

def momentum_coefficient(dot_m_b, V_b, U_inf, Ac):
    '''
    Calculate the momentum coefficient for boundary layer blowing.
    :param dot_m_b: Mass flow rate of the blown air.
    :param V_b: Blowing velocity.
    :param U_inf: Free stream velocity.
    :param Ac: Cross-sectional area affected by blowing.
    :return: Momentum coefficient.
    '''
    return (dot_m_b * V_b) / (0.5 * rho * U_inf**2 * Ac)

def drag_coefficient(C_D_0, delta_C_D_f):
    '''
    Calculate the overall drag coefficient with boundary layer
    ↪ control modifications.
    :param C_D_0: Baseline drag coefficient.
    :param delta_C_D_f: Change in drag coefficient due to skin
    ↪ friction reduction.
    :return: Total drag coefficient.
    '''
    return C_D_0 + delta_C_D_f

# Demo values
Re = 1e6  # Example Reynolds number
x = 1.0  # Distance from leading edge in meters
C_f = 0.005  # Example skin friction coefficient
phi = 0.1  # Effectiveness of surface treatment
Re_critical = 5e5  # Critical Reynolds number for transition
rho = 1.225  # Fluid density, kg/m^3 (air at sea level)
L = 1.0  # Characteristic length in meters
mu = 1.81e-5  # Dynamic viscosity, kg/(m·s) (air at sea level)
dot_m_b = 0.02  # Mass flow rate of blown air, kg/s
V_b = 10.0  # Blowing velocity, m/s
U_inf = 50.0  # Free stream velocity, m/s
Ac = 0.1  # Affected cross-sectional area, m^2
C_D_0 = 0.3  # Baseline drag coefficient
delta_C_D_f = -0.02  # Change in drag coefficient

# Calculations
delta = boundary_layer_thickness(Re, x)
delta_tau_w = skin_friction_reduction(C_f, phi)
v_s = suction_velocity(Re_critical, rho, L, mu)
C_mu = momentum_coefficient(dot_m_b, V_b, U_inf, Ac)
C_D = drag_coefficient(C_D_0, delta_C_D_f)

print("Boundary Layer Thickness (delta):", delta)
print("Change in Wall Shear Stress (delta_tau_w):", delta_tau_w)
print("Suction Velocity (v_s):", v_s)
print("Momentum Coefficient (C_mu):", C_mu)
print("Total Drag Coefficient (C_D):", C_D)
```

This code defines several key functions related to boundary layer control in rocket design:

- `boundary_layer_thickness` function computes the thickness of the boundary layer using Reynolds number and distance along the surface.

- `skin_friction_reduction` calculates the reduction in wall shear stress due to surface treatments designed to alter turbulence structures.

- `suction_velocity` determines the suction velocity necessary to control the boundary layer by removing low-momentum fluid.

- `momentum_coefficient` computes the coefficient used in boundary layer blowing techniques to energize the boundary layer and prevent separation.

- `drag_coefficient` calculates the overall drag coefficient factoring in modifications from boundary layer control to reduce skin friction drag.

The final block provides examples of these calculations with representative numerical values.

Chapter 59

Spacecraft Rendezvous and Docking Algorithms

Rendezvous Basics

Rendezvous operations in space involve complex orbital dynamics that require precise calculations to align trajectory vectors between two spacecrafts. The core of the rendezvous problem is solving the Lambert problem, which calculates the orbital transfer trajectory between two points at specified times. The time-of-flight t_f is critical and governed by the equation:

$$\Delta\theta = \int_{t_0}^{t_f} \frac{d\theta}{dt} dt$$

where $\Delta\theta$ symbolizes the phase angle change, and $\frac{d\theta}{dt}$ represents the time derivative of the true anomaly.

Clohessy-Wiltshire Equations

Near-Earth rendezvous typically employ the Clohessy-Wiltshire equations, which linearize the motion of a chaser vehicle relative to a target in a circular orbit. These equations are given by:

$$\ddot{x} - 3n^2x - 2n\dot{y} = 0,$$
$$\ddot{y} + 2n\dot{x} = 0,$$
$$\ddot{z} + n^2z = 0,$$

where n is the orbital mean motion, and (x, y, z) are the relative position coordinates. The solutions to these equations predict the trajectory corrections necessary for successful rendezvous maneuvers.

Phasing Maneuvers

Phasing involves adjusting the orbital period of a spacecraft to align its position with that of another. The delta-V (ΔV) required for a phasing maneuver can be determined by the vis-viva equation:

$$\Delta V = \sqrt{\mu\left(\frac{2}{r} - \frac{1}{a}\right)} - \sqrt{\mu\left(\frac{2}{r} - \frac{1}{a_{\text{desired}}}\right)}$$

where μ is the standard gravitational parameter, r is the radius at the burn point, and a is the semi-major axis of the initial orbit.

Docking Dynamics

Docking processes are governed by soft capture mechanisms that rely on precise control algorithms to ensure a smooth connection. Force and moment balance is crucial, where the docking force $\mathbf{F}$ and coupling torque $\mathbf{T}$ are expressed as:

$$\mathbf{F} = -k(\mathbf{x} - \mathbf{x}_d) - c\dot{\mathbf{x}}$$
$$\mathbf{T} = -k_\theta(\boldsymbol{\theta} - \boldsymbol{\theta}_d) - c_\theta\dot{\boldsymbol{\theta}}$$

where k and c are the linear structural stiffness and damping coefficients, $\mathbf{x}$ and $\boldsymbol{\theta}$ are the position and orientation vectors, and $\mathbf{x}_d$ and $\boldsymbol{\theta}_d$ denote the desired docking state.

Optimal Control in Rendezvous

Optimal control techniques, such as linear quadratic regulators (LQR), optimize the energy expenditures during the rendezvous by minimizing a performance index J. This is defined as:

$$J = \int_{t_0}^{t_f} \left(\mathbf{x}^T \mathbf{Q} \mathbf{x} + \mathbf{u}^T \mathbf{R} \mathbf{u} \right) dt$$

where **Q** and **R** are weight matrices corresponding to state and control efforts. **u** represents control input vectors, optimized by the LQR to achieve efficient trajectory corrections.

Spacecraft rendezvous and docking require precise coordination of orbital mechanics, trajectory adjustments, and dynamic control algorithms to ensure synchronization and safe coupling of vehicles in the microgravity environment of space.

Python Code Snippet

Below is a Python code snippet that encompasses the core computational elements for spacecraft rendezvous and docking algorithms, including solving the Lambert problem, Clohessy-Wiltshire equations, phasing maneuvers, docking dynamics, and optimal control techniques using Linear Quadratic Regulators (LQR).

```python
import numpy as np
from scipy.integrate import quad
from scipy.linalg import solve_continuous_are
from scipy.optimize import minimize


def solve_lambert(r1, r2, tf):
    '''
    Solve the Lambert problem to calculate the orbital transfer
    ↪   trajectory.
    :param r1: Initial position vector.
    :param r2: Final position vector.
    :param tf: Time of flight.
    :return: Transfer velocity vectors.
    '''
    # Placeholder for Lambert solver logic
    # Typically involves solving Lambert's equation
    pass


def clohessy_wiltshire_rel_motion(n, x0, y0, z0, vx0, vy0, vz0, t):
    '''
    Solve the Clohessy-Wiltshire equations for relative motion.
    :param n: Orbital mean motion.
    :param x0, y0, z0: Initial relative position.
    :param vx0, vy0, vz0: Initial relative velocity.
    :param t: Time elapsed.
```

```python
    :return: Relative position and velocity at time t.
    '''
    x = (4*x0 + 2*vy0/n - 3*x0*np.cos(n*t) + (2/n)*vy0*np.sin(n*t) +
    ↪    \
        ((6*x0*n + 2*vy0)*t - 3*x0 + 2*vy0/n)*np.cos(n*t)) / 4
    y = y0*np.cos(n*t) + (vx0/n)*np.sin(n*t)
    z = z0*np.cos(n*t) + (vz0/n)*np.sin(n*t)

    return x, y, z


def vis_viva(mu, r, a_initial, a_desired):
    '''
    Calculate delta-V for a phasing maneuver using vis-viva
    ↪  equation.
    :param mu: Gravitational parameter.
    :param r: Radius at burn point.
    :param a_initial: Initial semi-major axis.
    :param a_desired: Desired semi-major axis.
    :return: Delta-V required for the maneuver.
    '''
    return np.sqrt(mu * (2/r - 1/a_desired)) - np.sqrt(mu * (2/r -
    ↪  1/a_initial))


def docking_forces(x, xd, theta, thetad, k, c, k_theta, c_theta,
↪   x_dot, theta_dot):
    '''
    Calculate docking forces and torques.
    :param x: Current position.
    :param xd: Desired docking position.
    :param theta: Current orientation.
    :param thetad: Desired docking orientation.
    :param k: Linear stiffness coefficient.
    :param c: Linear damping coefficient.
    :param k_theta: Angular stiffness coefficient.
    :param c_theta: Angular damping coefficient.
    :param x_dot: Current docking velocity.
    :param theta_dot: Current angular velocity.
    :return: Force and torque for docking.
    '''
    force = -k * (x - xd) - c * x_dot
    torque = -k_theta * (theta - thetad) - c_theta * theta_dot

    return force, torque


def lqr_control(A, B, Q, R):
    '''
    Computes the optimal state feedback LQR controller gain.
    :param A: System state matrix.
    :param B: Control input matrix.
    :param Q: State weighting matrix.
```

```
    :param R: Control input weighting matrix.
    :return: Optimal gain matrix K.
    '''
    P = solve_continuous_are(A, B, Q, R)
    K = np.linalg.inv(R) @ B.T @ P
    return K


# Example parameters for demonstration purposes
mu_earth = 398600   # Gravitational parameter (km^3/s^2)
n_example = 0.001   # Example mean-motion value
x0, y0, z0 = 0, 0, 0
vx0, vy0, vz0 = 0.1, 0.1, 0.1
t_example = 100   # Time elapsed (s)
r_example = 7000   # Example orbit radius

# Placeholder for system dynamics matrices
A_example = np.array([[0, 1], [0, 0]])
B_example = np.array([[0], [1]])
Q_example = np.eye(2)
R_example = np.array([[1]])

x, y, z = clohessy_wiltshire_rel_motion(n_example, x0, y0, z0, vx0,
    ↪  vy0, vz0, t_example)
delta_v = vis_viva(mu_earth, r_example, 6700, 6800)
force, torque = docking_forces(x, 0, 0, 0, 100, 10, 50, 5, vx0, 0)
K_lqr = lqr_control(A_example, B_example, Q_example, R_example)

print("Relative Position:", x, y, z)
print("Delta V:", delta_v)
print("Docking Force:", force, "Docking Torque:", torque)
print("LQR Gain Matrix:", K_lqr)
```

This code defines several key functions necessary for implementing core spacecraft rendezvous and docking operational strategies:

- solve_lambert is a placeholder that would solve the Lambert problem using appropriate algorithms.

- clohessy_wiltshire_rel_motion solves relative motion for near-Earth objects using Clohessy-Wiltshire equations.

- vis_viva calculates the required delta-V for phasing maneuvers, leveraging the vis-viva equation.

- docking_forces determines the forces and torques required for effective docking.

- lqr_control computes the optimal gain matrix for an LQR problem, enabling energy-efficient control.

The code example shows how to use these functions with sample data, demonstrating real-world applications of these fundamental concepts.

Chapter 60

Leaky Integrations in Control Systems

Introduction to Leaky Integrators

Leaky integrators represent a cornerstone in control system design, particularly in aerospace applications where robust stability and responsiveness are paramount. These components act to mitigate the potential for integrator windup by incorporating a leakage factor into the integrative process. The mathematical representation of a leaky integrator can be described by the equation:

$$I(s) = \frac{1}{\tau s + 1}$$

where τ is the leakage time constant, and s is the Laplace transform variable. The introduction of the leakage factor τ provides a controlled reduction in the accumulation of error by gradually dissipating the stored integral value, thus enhancing the system's overall stability.

Feedback Loop Design with Leaky Integrators

Integrating leaky integrators into feedback loops addresses the dual challenges of maintaining system responsiveness and damping oscillations. The feedback transfer function incorporating a leaky integrator is given by:

$$G(s) = \frac{K_p + K_i \frac{1}{\tau s+1} + K_d s}{1 + G(s)H(s)}$$

here, K_p, K_i, and K_d represent the proportional, integral, and derivative gains respectively, while $H(s)$ symbolizes the feedback path. The inclusion of the leaky integrator term, $\frac{K_i}{\tau s+1}$, ensures that the integrated error does not grow unbounded, preserving control authority and preventing excessive actuator wear, which is critical in aerospace systems.

Stability Analysis Through Eigenvalue Evaluation

Evaluating the stability properties of a control system with leaky integrators involves analyzing the system's eigenvalues. The characteristic polynomial of the closed-loop system must be derived and solved to determine the eigenvalues' location in the complex plane. The roots of the polynomial given by:

$$P(s) = 1 + G(s)H(s)$$

depend on the leakage factor τ and system gains, dictating the damping ratio and natural frequency. For a system governed by second-order dynamics, the adjusted pole positions can be expressed as:

$$s_{1,2} = -\frac{\omega_n \zeta}{\tau} \pm \omega_n \sqrt{1 - \zeta^2}$$

where ω_n is the natural frequency and ζ is the damping ratio. Proper tuning of τ is required to prevent the poles from crossing into the right half of the complex s-plane, which would imply instability in real-world applications.

Implementation in Rocket Attitude Control Systems

The application of leaky integrators in rocket attitude control systems enhances the control feedback loop's ability to manage disturbances and perturbations effectively. Attitude control systems often rely on gyroscopic sensors that can integrate accumulated errors over time. A typical representation in state-space form is:

$$\dot{\mathbf{x}} = \mathbf{A}\mathbf{x} + \mathbf{B}\text{LeakyIntegrator}(\mathbf{y})$$

where **x** denotes the state vector, and **y** represents the sensor output subject to integration. By replacing the pure integration block with a leaky integrator, the system exhibits increased tolerance to noise and unexpected angular momentum changes, which are prevalent during rocket operations.

Sensitivity Functions and Robustness

The sensitivity function $S(s)$ offers insight into the robustness of systems utilizing leaky integrators. It is defined as:

$$S(s) = \frac{1}{1 + G(s)H(s)}$$

By including leakage in the integration pathway, the peak magnitude of $S(s)$ is attenuated, reducing vulnerability to gain variations. This characteristic is essential for ensuring robustness in the uncertain and dynamically shifting environments typical of space travel.

Leaky integrators thus play a critical role in refining the responsiveness and stability of rocket control systems through carefully adjusted damping, contributing significantly to the advancement of aerospace control dynamics.

Python Code Snippet

Below is a Python code snippet that encompasses the implementation of key equations and algorithms related to leaky integrators and their application in control systems, particularly within the context of rocket attitude control.

```python
import numpy as np
import scipy.signal as signal
import matplotlib.pyplot as plt

def leaky_integrator_parameters(tau, s):
    '''
    Return the transfer function of a leaky integrator.
```

```
    :param tau: Leakage time constant.
    :param s: Laplace variable (complex frequency).
    :return: Transfer function.
    '''
    return 1 / (tau * s + 1)

def feedback_loop(kp, ki, kd, tau, g_s, h_s):
    '''
    Calculate feedback loop transfer function with leaky integrator.
    :param kp: Proportional gain.
    :param ki: Integral gain.
    :param kd: Derivative gain.
    :param tau: Leakage time constant.
    :param g_s: Open loop gain.
    :param h_s: Feedback path gain.
    :return: Feedback transfer function.
    '''
    leaky_integrator_tf = ki * leaky_integrator_parameters(tau,
    ↪   signal.TransferFunction.s)
    numerator = kp + leaky_integrator_tf + kd *
    ↪   signal.TransferFunction.s
    feedback_tf = signal.TransferFunction(numerator.num,
    ↪   numerator.den) / (1 + g_s * h_s)
    return feedback_tf

def stability_analysis(tau, omega_n, zeta):
    '''
    Analyze stability through eigenvalue evaluation.
    :param tau: Leakage time constant.
    :param omega_n: Natural frequency.
    :param zeta: Damping ratio.
    :return: Eigenvalues (pole positions).
    '''
    roots = np.roots([1, 2*zeta*omega_n/tau, omega_n**2])
    return roots

def simulate_attitude_control(tau, A, B, y_initial):
    '''
    Simulate attitude control system using leaky integrators.
    :param tau: Leakage time constant.
    :param A: State matrix.
    :param B: Input matrix.
    :param y_initial: Initial sensor output.
    :return: Time series of system states.
    '''
    leaky_output = lambda y: signal.lfilter([1], [tau, 1], y)
    t = np.linspace(0, 10, 1000)
    y = leaky_output(y_initial)
    _, y_out, _ = signal.lsim((A, B), u=y, T=t)
    return t, y_out

def robustness_metrics(g_s, h_s):
    '''
```

```
    Calculate sensitivity function to assess robustness.
    :param g_s: Open loop gain.
    :param h_s: Feedback path gain.
    :return: Sensitivity function magnitude.
    '''
    sensitivity = 1 / (1 + g_s * h_s)
    _, mag, _ = signal.bode(sensitivity)
    return mag

# Example calculations
tau_example = 0.5
s_example = signal.TransferFunction.s
leaky_integrator_tf_example =
↪   leaky_integrator_parameters(tau_example, s_example)
feedback_tf_example = feedback_loop(1.0, 0.5, 0.1, tau_example,
↪   leaky_integrator_tf_example, 1.0)
roots_example = stability_analysis(tau_example, 5, 0.7)

print("Leaky Integrator Transfer Function:",
↪   leaky_integrator_tf_example)
print("Feedback Transfer Function:", feedback_tf_example)
print("Stability Eigenvalues:", roots_example)

# Plotting a simulation
A_example = np.array([[-0.5, 0.1], [0.2, -0.3]])
B_example = np.array([1.0, 0.0])
y_initial_example = np.array([0.0, 1.0])

t_sim, y_sim = simulate_attitude_control(tau_example, A_example,
↪   B_example, y_initial_example)

plt.plot(t_sim, y_sim)
plt.title("Attitude Control Simulation with Leaky Integrators")
plt.xlabel("Time")
plt.ylabel("System Response")
plt.grid()
plt.show()

# Evaluate robustness
robustness_mag = robustness_metrics(feedback_tf_example.num,
↪   feedback_tf_example.den)
plt.plot(robustness_mag)
plt.title("Sensitivity Function Magnitude")
plt.xlabel("Frequency")
plt.ylabel("Magnitude (dB)")
plt.grid()
plt.show()
```

This code defines several key functionalities necessary for understanding and applying leaky integrators in control systems:

- `leaky_integrator_parameters` implements the transfer function of a leaky integrator.

- `feedback_loop` calculates a feedback loop transfer function incorporating a leaky integrator.

- `stability_analysis` evaluates the stability of the control system by finding eigenvalues.

- `simulate_attitude_control` simulates a rocket attitude control system using state-space representation and leaky integrators.

- `robustness_metrics` assesses system robustness using the sensitivity function.

The final block of code provides an example of these calculations and visualizations, demonstrating the practical application of these concepts in aerospace control systems.

Chapter 61

Limitations of Linear Control Techniques

Linear Control Approaches in Aerospace Engineering

Linear control systems have traditionally played a pivotal role in aerospace applications, offering simplicity and predictability. Fundamental to these systems is the assumption that the response of the system is directly proportional to the input, enabling linear controllers to manage a wide array of aerospace vehicles with acceptable performance. The general framework for linear control can be defined by state-space representation:

$$\dot{\mathbf{x}} = \mathbf{A}\mathbf{x} + \mathbf{B}\mathbf{u}$$

$$\mathbf{y} = \mathbf{C}\mathbf{x} + \mathbf{D}\mathbf{u}$$

where $\mathbf{x}$ is the state vector, $\mathbf{u}$ is the control input, and $\mathbf{y}$ is the output. The matrices $\mathbf{A}$, $\mathbf{B}$, $\mathbf{C}$, and $\mathbf{D}$ define the system dynamics and outputs.

Challenges in Dynamic Rocket Systems

In highly dynamic rocket systems, linear control techniques encounter several critical challenges. Nonlinearities inherent in aerodynamic forces, propulsion systems, and structural flexibilities pose significant obstacles. The dynamics of a rocket in varying atmospheric conditions and during different phases of flight entail substantial deviations from linear behavior.

Consider the control input to thrust relationship expressed by:

$$F = m \cdot \dot{v} = T - D - mg$$

where F is the net force, m is the mass, T is thrust, D is drag, and g is gravitational acceleration. Linear controllers often assume constant coefficients, lacking adaptability to nonlinear changes such as variations in air density and fuel burn rates.

Stability and Robustness Limitations

Linear control systems inherently possess limitations in stability and robustness when applied to rockets under rapid or unpredictable environmental changes. The small-gain theorem for assessing robust performance may not encompass the system's behavior under all conditions:

$$\|T(s) - 1\| < 1 \implies \|S(s)\| < 1$$

where $T(s)$ is the complementary sensitivity function and $S(s)$ is the sensitivity function.

Unanticipated disturbances and parameter uncertainties exacerbate the instability in linear controllers. For instance, a fixed-gain PID controller might struggle to sustain stability due to resonant frequencies in flexible modes, a scenario frequent in elongated rocket fuselages.

Transition to Non-Linear Control Methods

As the controllability of linear systems reaches its limits in complex aerospace scenarios, transitioning to nonlinear control methodologies becomes imperative. Nonlinear controllers accommodate system behavior more accurately by leveraging models that reflect real-world dynamics, like feedback linearization and backstepping techniques.

A nonlinear state feedback can be expressed as:

$$\mathbf{u} = \mathbf{k}(\mathbf{x}) + \mathbf{v}$$

where **k** represents the nonlinear state feedback law, incorporating differential geometry insight.

Lyapunov's direct method provides a framework:

$$V(\mathbf{x}) = \mathbf{x}^T \mathbf{P} \mathbf{x}$$

$$\dot{V}(\mathbf{x}) = \mathbf{x}^T (\mathbf{A}^T \mathbf{P} + \mathbf{P} \mathbf{A}) \mathbf{x} < 0$$

where V is a Lyapunov candidate function, ensuring stability through negative definiteness.

Model Predictive Control and Adaptive Techniques

Erosion of linear control assumptions advocates for model predictive control (MPC) and adaptive control strategies. MPC recasts control problems within a finite future window, optimizing control moves via quadratic programming or iterative solvers subject to constraints:

$$\min_{\mathbf{u}} \sum_{k=0}^{N} \|\mathbf{y}(k) - \mathbf{r}(k)\|_Q^2 + \|\mathbf{u}(k)\|_R^2$$

Adaptive techniques, like gain-scheduled control and self-tuning regulators, adapt parameters in real-time, accommodating variations in system dynamics without a priory established limits or fixed linear models.

This intrinsic flexibility propels nonlinear methods as quintessential to meeting the challenges posed by the complex, multi-faceted nature of modern aerospace applications, particularly in dynamic rocket systems.

Python Code Snippet

Below is a Python code snippet that encompasses the core computational elements of linear and nonlinear control system equations, state-space representation, and some aspects of model predictive control and Lyapunov's method.

```python
import numpy as np
from scipy.linalg import solve_continuous_are
from scipy.optimize import minimize

# Linear system matrices (example values)
A = np.array([[0.0, 1.0], [-1.0, -1.0]])
B = np.array([[0.0], [1.0]])
C = np.array([[1.0, 0.0]])
D = np.array([[0.0]])

# State-space system dynamics: dx/dt = A * x + B * u
def linear_state_space(x, u):
    dxdt = A @ x + B @ u
    return dxdt

# Non-linear control (feedback linearization example)
def nonlinear_state_feedback(x):
    u = np.sin(x[0]) + 0.5 * x[1]  # A simple nonlinear feedback law
    return u

# Lyapunov function and its derivative
def lyapunov_function(x, P):
    return x.T @ P @ x

def derivative_lyapunov(x, A, P):
    return x.T @ (A.T @ P + P @ A) @ x

# Example of a nonlinear feedback control application
def simulate_nonlinear_control(x0, time, dt):
    x = np.array(x0)
    trajectory = [x]
    P = np.eye(len(x0))  # Example P matrix for Lyapunov function
    for _ in range(int(time / dt)):
        u = nonlinear_state_feedback(x)
        dxdt = linear_state_space(x, u)
        x = x + dxdt * dt
        V = lyapunov_function(x, P)
        dV = derivative_lyapunov(x, A, P)
        assert dV < 0, "System is not stable"
        trajectory.append(x)
    return np.array(trajectory)

# An example of Model Predictive Control (MPC)
def mpc_control(x0, horizon, constraints):
    def objective(u_sequence):
        cost = 0
        x = np.array(x0)
        for u in u_sequence:
            dxdt = linear_state_space(x, u)
            x = x + dxdt * 0.1  # Assume a time step of 0.1
            cost += x.T @ x + u.T @ u
        return cost
```

```
u0 = [0] * horizon
result = minimize(objective, u0, constraints=constraints)
return result.x

# Example constraints for MPC
constraints = [{'type': 'ineq', 'fun': lambda u: u[-1] - 1}]  #
↪ Example: Last control input less than 1

# Example initial condition
x0 = [1.0, 0.0]

# Running simulations
trajectory = simulate_nonlinear_control(x0, 10, 0.1)
mpc_solution = mpc_control(x0, 20, constraints)

# Results
print("Nonlinear Control Trajectory:", trajectory)
print("MPC Solution:", mpc_solution)
```

This code defines several key functions necessary for the exploration of linear and nonlinear control techniques:

- `linear_state_space` function simulates the state-space representation for linear control systems based on matrices **A** and **B**.

- `nonlinear_state_feedback` provides a simple illustration of nonlinear feedback control by applying a custom feedback law.

- `lyapunov_function` and its derivative are crucial for ensuring system stability via Lyapunov's method.

- `simulate_nonlinear_control` simulates a control trajectory by applying a nonlinear feedback law over time.

- `mpc_control` presents a basic setup for Model Predictive Control, optimizing control inputs over a horizon subject to given constraints.

The final block of code simulates both a nonlinear control trajectory and solves an MPC problem, demonstrating the application of these techniques in dynamic aerospace systems.

Chapter 62

Genetic Algorithms for Optimizing Propulsion

Introduction to Genetic Algorithms

Genetic algorithms (GAs) offer robust optimization capabilities by mimicking natural evolutionary processes. Integral to aerospace propulsion systems, the algorithm iteratively evolves a population of potential solutions, evaluating, selecting, and refining candidate designs based on a fitness function. Solutions are encoded as chromosomes, often binary strings or real-valued vectors, representing different propulsion system designs.

Chromosome Representation and Fitness Evaluation

In aerospace propulsion optimization, the representation of candidate solutions, also known as chromosome encoding, should accurately capture design parameters and constraints. Each chromosome **c** encompasses variables such as nozzle shape, fuel distribution, and material properties, defining the design space:

$$\mathbf{c} = [c_1, c_2, \ldots, c_n]$$

where c_i are individual design variables. The fitness function $f(\mathbf{c})$ quantitatively evaluates propulsion efficiency and performance.

Key performance metrics include specific impulse I_{sp}, thrust T, and efficiency η:

$$f(\mathbf{c}) = w_1 \cdot I_{sp}(\mathbf{c}) + w_2 \cdot T(\mathbf{c}) + w_3 \cdot \eta(\mathbf{c})$$

Weights w_1, w_2, w_3 prioritize objectives according to mission requirements.

Selection Mechanisms and Genetic Operators

The selection process determines which chromosomes progress to subsequent generations. Techniques such as roulette wheel selection and tournament selection ensure fitter individuals have a higher probability of contributing to progeny. The selection probability P_s for a chromosome depends on its relative fitness:

$$P_s(c_i) = \frac{f(c_i)}{\sum_{j=1}^{N} f(c_j)}$$

Genetic operators—crossover and mutation—introduce diversity and facilitate exploration of the design space. Crossover combines segments of parent chromosomes to produce new offspring:

$$\mathbf{c}'_{off1} = [c_{p1,1}, \ldots, c_{p1,k}, c_{p2,k+1}, \ldots, c_{p2,n}]$$
$$\mathbf{c}'_{off2} = [c_{p2,1}, \ldots, c_{p2,k}, c_{p1,k+1}, \ldots, c_{p1,n}]$$

Mutation introduces random alterations, mitigating premature convergence:

$$c'_i = \begin{cases} \texttt{random\_value()}, & \text{with probability } p_{mut} \\ c_i, & \text{otherwise} \end{cases}$$

Convergence and Computational Considerations

Convergence to optimal solutions hinges on appropriately tuning parameters such as population size, crossover rate p_{cross}, and mutation rate p_{mut}. Genetic algorithms balance exploration and exploitation, necessitating empirical or adaptive tuning strategies.

The computational expense of GAs depends on fitness evaluations, often dominated by complex propulsion simulations. Efficient implementation may leverage parallelism or surrogate models to accelerate evaluations while preserving solution quality.

Application in Propulsion System Design

In optimizing propulsion systems, genetic algorithms critically assess trade-offs in design parameters impacting thrust efficiency and resource utilization. The iterative algorithmic process explores configurations of components, such as turbopumps and combustion chambers, advancing design towards a global optimum.

The robustness of genetic algorithms in handling complex, multi-objective optimization problems underscores their applicability across a spectrum of aerospace propulsion challenges, leading to significant advancements in propulsion technologies.

Python Code Snippet

Below is a Python code snippet that encompasses the critical computational elements necessary for implementing genetic algorithms in propulsion system design optimization, including functions for chromosome encoding, fitness evaluation, selection, crossover, and mutation.

```
import numpy as np
import random

def fitness_function(chromosome):
    '''
    Calculate the fitness of a chromosome based on propulsion
        metrics.
    :param chromosome: List of design variables [c1, c2, ..., cn].
    :return: Fitness value.
    '''
    I_sp = calculate_specific_impulse(chromosome)
    T = calculate_thrust(chromosome)
    efficiency = calculate_efficiency(chromosome)

    # Weights for each component of the fitness function
    w1, w2, w3 = 0.5, 0.3, 0.2

    return w1 * I_sp + w2 * T + w3 * efficiency
```

```python
def roulette_wheel_selection(population, fitness_values):
    '''
    Select a chromosome from the population based on roulette wheel
    ↪ selection.
    :param population: List of chromosomes.
    :param fitness_values: List of fitness values corresponding to
    ↪ the population.
    :return: Selected chromosome.
    '''
    total_fitness = sum(fitness_values)
    selection_probs = [f / total_fitness for f in fitness_values]
    return population[np.random.choice(len(population),
    ↪ p=selection_probs)]

def crossover(parent1, parent2):
    '''
    Perform crossover operation between two parent chromosomes.
    :param parent1: First parent chromosome.
    :param parent2: Second parent chromosome.
    :return: Two offspring chromosomes.
    '''
    point = random.randint(1, len(parent1) - 1)
    offspring1 = parent1[:point] + parent2[point:]
    offspring2 = parent2[:point] + parent1[point:]
    return offspring1, offspring2

def mutate(chromosome, mutation_rate=0.01):
    '''
    Perform mutation operation on a chromosome.
    :param chromosome: Chromosome to mutate.
    :param mutation_rate: Probability of mutation for each gene.
    :return: Mutated chromosome.
    '''
    return [
        element if random.random() > mutation_rate else
        ↪ random_value()
        for element in chromosome
    ]

def optimize(num_generations=100, population_size=50):
    '''
    Run genetic algorithm for optimizing propulsion systems.
    :param num_generations: Number of generations to evolve.
    :param population_size: Size of the population.
    :return: Best solution found.
    '''
    # Initialize population
    population = [random_chromosome() for _ in
    ↪ range(population_size)]
    for generation in range(num_generations):
        fitness_values = [fitness_function(chromosome) for
        ↪ chromosome in population]
        new_population = []
```

```python
        for _ in range(population_size // 2):
            parent1 = roulette_wheel_selection(population,
             ↪  fitness_values)
            parent2 = roulette_wheel_selection(population,
             ↪  fitness_values)
            offspring1, offspring2 = crossover(parent1, parent2)
            new_population.extend([mutate(offspring1),
             ↪  mutate(offspring2)])
        population = new_population

        if generation % 10 == 0:
            print(f"Generation {generation}: Best Fitness =
             ↪  {max(fitness_values)}")

    best_solution = max(population, key=fitness_function)
    return best_solution

def random_chromosome():
    '''
    Generate a random chromosome representing a propulsion system
     ↪  design.
    :return: Randomly generated chromosome.
    '''
    # Example implementation; customize as needed for the problem
    return [random_value() for _ in range(10)]

def random_value():
    '''
    Generate a random value within a specified range.
    :return: Randomly generated value.
    '''
    return random.uniform(0, 1)

# Placeholder functions for propulsion calculations
def calculate_specific_impulse(chromosome):
    ''' Example placeholder function '''
    return sum(chromosome)  # Replace with actual calculation

def calculate_thrust(chromosome):
    ''' Example placeholder function '''
    return sum(chromosome)  # Replace with actual calculation

def calculate_efficiency(chromosome):
    ''' Example placeholder function '''
    return sum(chromosome) / len(chromosome)  # Replace with actual
     ↪  calculation

# Run the optimization
best_solution = optimize()
print("Best Chromosome Found:", best_solution)
```

In this code snippet, several critical functions are defined to implement genetic algorithms for optimizing aerospace propulsion systems:

- `fitness_function` evaluates the effectiveness of a design based on propulsion performance metrics.

- `roulette_wheel_selection` implements a selection strategy to favor more fit chromosomes.

- `crossover` combines two parent chromosomes to generate offspring.

- `mutate` introduces random changes to a chromosome to explore the solution space.

- `optimize` orchestrates the genetic algorithm process, iterating through generations to find an optimal solution.

- Additional placeholder functions (`calculate_specific_impulse`, `calculate_thrust`, `calculate_efficiency`) and helper functions (`random_chromosome`, `random_value`) are provided to complete the implementation.

The final block runs the optimization to find the best design solution based on the defined fitness landscape.

Chapter 63

Data Fusion in Multi-Sensor Navigation Systems

Introduction to Multi-Sensor Data Fusion

Multi-sensor data fusion in navigation systems leverages the complementary characteristics of various sensor modalities to enhance overall system accuracy and robustness. In autonomous space missions, precise navigation is imperative due to the absence of traditional terrestrial navigation aids. Data fusion methodologies, like the Kalman Filter and its derivatives, play a vital role in synthesizing data from sources such as inertial measurement units (IMUs), star trackers, and radio navigation systems.

Kalman Filter for Data Fusion

The `Kalman Filter` is a widely adopted algorithm for linear data fusion in navigation systems, providing optimal estimates of the system's state vector x_k by minimizing the error covariance. The recursive process involves prediction and update steps. In the prediction phase:

$$\mathbf{x}_{k|k-1} = \mathbf{F}_k \mathbf{x}_{k-1|k-1} + \mathbf{B}_k \mathbf{u}_k \qquad (63.1)$$

$$\mathbf{P}_{k|k-1} = \mathbf{F}_k \mathbf{P}_{k-1|k-1} \mathbf{F}_k^T + \mathbf{Q}_k \qquad (63.2)$$

where $\mathbf{F}_k$ is the state transition model, $\mathbf{B}_k$ is the control input model, $\mathbf{u}_k$ represents control inputs, and $\mathbf{Q}_k$ is process noise covariance.

During the update phase:

$$\mathbf{K}_k = \mathbf{P}_{k|k-1} \mathbf{H}_k^T (\mathbf{H}_k \mathbf{P}_{k|k-1} \mathbf{H}_k^T + \mathbf{R}_k)^{-1} \qquad (63.3)$$

$$\mathbf{x}_{k|k} = \mathbf{x}_{k|k-1} + \mathbf{K}_k (\mathbf{z}_k - \mathbf{H}_k \mathbf{x}_{k|k-1}) \qquad (63.4)$$

$$\mathbf{P}_{k|k} = (\mathbf{I} - \mathbf{K}_k \mathbf{H}_k) \mathbf{P}_{k|k-1} \qquad (63.5)$$

Here, $\mathbf{H}_k$ is the observation model, $\mathbf{R}_k$ is the observation noise covariance, $\mathbf{K}_k$ is the Kalman gain, and $\mathbf{z}_k$ represents sensor measurements.

Nonlinear Filtering Techniques

Nonlinear systems, common in aerospace applications, require augmented filtering approaches such as the `Extended Kalman Filter` (EKF) and `Unscented Kalman Filter` (UKF). The EKF linearizes the system dynamics around current estimates:

$$\mathbf{F}_k = \left. \frac{\partial \mathbf{f}}{\partial \mathbf{x}} \right|_{\mathbf{x} = \mathbf{x}_{k-1|k-1}} \qquad (63.6)$$

$$\mathbf{H}_k = \left. \frac{\partial \mathbf{h}}{\partial \mathbf{x}} \right|_{\mathbf{x} = \mathbf{x}_{k|k-1}} \qquad (63.7)$$

where $\mathbf{f}$ and $\mathbf{h}$ are nonlinear functions describing the process and measurement models, respectively.

The UKF implements a deterministic sampling approach to capture the mean and covariance more accurately through a set of sigma points. This eliminates the need for explicit linearization and enhances performance in systems exhibiting marked nonlinearity.

Implementing Data Fusion for Enhanced Navigation Accuracy

In autonomous space missions, integrating data from sensors such as accelerometers, gyroscopes, magnetometers, and celestial navigation instruments necessitates a synergistic fusion strategy. The mathematical model underpinning this integration can be articulated as:

$$\mathbf{z}_k = \mathbf{h}(\mathbf{x}_k) + \mathbf{v}_k \tag{63.8}$$

$$\mathbf{x}_{k+1} = \mathbf{f}(\mathbf{x}_k, \mathbf{u}_k) + \mathbf{w}_k \tag{63.9}$$

Where $\mathbf{v}_k$ and $\mathbf{w}_k$ denote measurement and process noise, assumed to follow a Gaussian distribution with zero mean.

These multi-sensor fusion techniques optimize the utilization of available sensor data by reconciling discrepancies and generating accurate state estimations, thereby realizing improved navigational precision crucial for executing intricate space maneuvers and ensuring mission success.

Python Code Snippet

Below is a Python code snippet that encompasses the core computational elements for implementing data fusion in multi-sensor navigation systems, including the Kalman Filter and its extended variants for nonlinear systems.

```
import numpy as np

def kalman_filter(F, B, H, Q, R, z, x_prior, P_prior, u=None):
    '''
    Basic Kalman filter implementation for fusion of sensor data.
    :param F: State transition model.
    :param B: Control input model.
    :param H: Observation model.
    :param Q: Process noise covariance.
    :param R: Measurement noise covariance.
    :param z: Measurement vector.
    :param x_prior: State estimate from previous time step.
    :param P_prior: Estimate covariance from previous time step.
    :param u: Control input vector, optional.
    :return: Updated state estimate and covariance.
    '''
```

```python
    # Prediction step
    x_pred = np.dot(F, x_prior) + np.dot(B, u) if u is not None else
    ↪ np.dot(F, x_prior)
    P_pred = np.dot(F, np.dot(P_prior, F.T)) + Q

    # Update step
    y = z - np.dot(H, x_pred)  # Innovation
    S = np.dot(H, np.dot(P_pred, H.T)) + R  # Innovation covariance
    K = np.dot(P_pred, np.dot(H.T, np.linalg.inv(S)))  # Kalman
    ↪ gain
    x_est = x_pred + np.dot(K, y)
    P_est = P_pred - np.dot(K, np.dot(H, P_pred))

    return x_est, P_est

def extended_kalman_filter(f, h, F_jacobian, H_jacobian, Q, R, z,
↪ x_prior, P_prior, u=None):
    '''
    Extended Kalman filter for nonlinear systems.
    :param f: Nonlinear state transition function.
    :param h: Nonlinear observation function.
    :param F_jacobian: Jacobian of the state transition function.
    :param H_jacobian: Jacobian of the observation function.
    :param Q: Process noise covariance.
    :param R: Measurement noise covariance.
    :param z: Measurement vector.
    :param x_prior: State estimate from previous time step.
    :param P_prior: Estimate covariance from previous time step.
    :param u: Control input vector, optional.
    :return: Updated state estimate and covariance.
    '''
    # Prediction step
    x_pred = f(x_prior, u) if u is not None else f(x_prior)
    F = F_jacobian(x_prior, u)
    P_pred = np.dot(F, np.dot(P_prior, F.T)) + Q

    # Update step
    H = H_jacobian(x_pred)
    y = z - h(x_pred)
    S = np.dot(H, np.dot(P_pred, H.T)) + R
    K = np.dot(P_pred, np.dot(H.T, np.linalg.inv(S)))
    x_est = x_pred + np.dot(K, y)
    P_est = P_pred - np.dot(K, np.dot(H, P_pred))

    return x_est, P_est

# Example usage with placeholder functions and dummy data
F = np.array([[1, 0], [0, 1]])  # Identity for simple constant
↪ velocity model
B = np.array([[0.5, 0], [0, 0.5]])  # Assuming control inputs are
↪ temporally halfway effective
H = np.array([[1, 0], [0, 1]])  # Direct observation model
Q = np.array([[0.1, 0], [0, 0.1]])  # Small process noise covariance
```

```
R = np.array([[1, 0], [0, 3]])   # Measurement noise with more
↪ uncertainty in second dimension
z = np.array([10, 20])   # Observed state
x_prior = np.array([9, 19])   # Previous state estimate
P_prior = np.eye(2)   # Previous estimate covariance (identity for
↪ simplicity)
u = np.array([0.1, 0.1])   # Control input vector

# Run Kalman Filter
x_est, P_est = kalman_filter(F, B, H, Q, R, z, x_prior, P_prior, u)

print("Estimated State:", x_est)
print("Estimated Covariance:", P_est)
```

This code defines two key filtering functions that are essential for multi-sensor data fusion:

- `kalman_filter` performs the basic linear Kalman filtering process, predicting and updating the state estimates based on sensor data.

- `extended_kalman_filter` extends the basic filter for nonlinear systems using jacobians of the system dynamics.

The example block shows how these functions could be applied using simplified system models and sensor data, demonstrating the process of state estimation critical for precise navigation in complex environments.

Chapter 64

Sounding Rockets and High-Altitude Research

Introduction

Sounding rockets serve as essential tools in aerospace research, facilitating high-altitude investigations where conventional aircraft and balloons are ineffective. These rockets are instrumental in gathering data on atmospheric properties, astrophysical phenomena, and microgravity conditions. The design requirements and operational constraints of sounding rockets are uniquely tailored to their missions, reflecting aspects such as altitude range, payload capacity, and cost efficiency.

Launch Dynamics and Altitude Achievements

The dynamics of sounding rocket launches are inherently transient, characterized by brief and high-acceleration phases. A critical parameter in this context is the altitude h attained, which can be modeled as follows:

$$h = \frac{v_0^2 \sin^2 \theta}{2g}$$

where v_0 denotes the initial velocity, θ is the launch angle, and g signifies gravitational acceleration. The altitude is further influenced by aerodynamic forces, particularly drag D, which is expressed in terms of the drag coefficient C_D, air density ρ, reference area A, and velocity v:

$$D = \frac{1}{2} C_D \rho A v^2$$

1 Payload Considerations

Payload capacity is a decisive factor in the design of sounding rockets, determined by the thrust-to-weight ratio T/W. This ratio impacts the maximum achievable altitude and payload safety during ascent. The thrust T can be calculated based on the exhaust velocity v_e and the mass flow rate $\dot{m}$:

$$T = \dot{m} v_e$$

Considerations must also account for structural constraints and vibration loads, which are determined through finite element analysis (FEA) to ensure payload integrity.

Trajectory and Guidance Systems

The trajectory path is governed by Newton's equations of motion under the influence of gravitational and aerodynamic forces. The fundamental equations for a two-dimensional, non-rotating Earth model are:

$$\frac{dv}{dt} = \frac{T - D}{m} - g \cos \theta$$

$$\frac{d\theta}{dt} = \frac{v \cos \theta}{R + h}$$

where m is the mass of the rocket, R is the Earth's radius, and the derivatives $\frac{dv}{dt}$ and $\frac{d\theta}{dt}$ represent the time rate of change of velocity and angle, respectively. Electronic guidance systems employing inertial navigation and GPS data logger integration provide real-time trajectory correction capabilities.

Aerodynamic and Thermal Constraints

Sounding rockets, during their brief but intense ascents, encounter significant aerodynamic heating. The convective heat transfer coefficient h_c is expressed as a function of the Nusselt number Nu:

$$Nu = \frac{h_c L}{k}$$

where L is the characteristic length and k is the thermal conductivity of the medium. To ensure structural integrity, heat shields and thermal coatings are applied, designed using computational fluid dynamics (CFD) simulations to predict and mitigate adverse thermal effects.

1 Material Selection and Fabrication

Material selection for the rocket's body and heat shields must account for high-temperature tolerance and mechanical strength. Metals such as titanium and high-strength alloys are commonly used, subject to verification through thermal cycling tests and stress-strain analyses.

Data Collection and Analysis

The data acquired by sounding rocket payloads is typically focused on atmospheric measurements including temperature, pressure, and chemical composition. These measurements are captured using a suite of on-board sensors and transmitted via telemetric systems for post-flight analysis. The data processing employs advanced statistical techniques to interpret environmental interactions and phenomena observed during the mission.

In sum, sounding rockets are versatile yet complex vehicles uniquely suited for high-altitude atmospheric research, necessitating meticulous engineering design and operational considerations to achieve mission objectives and ensure collected data reliability. The integration of contemporary computational tools and materials science advancements continues to enhance the effectiveness and reach of these exploratory platforms.

Python Code Snippet

Below is a Python code snippet that encompasses the core computational elements for modeling and analyzing sounding rocket dynamics and performance, including altitude calculations, aerodynamic drag assessment, trajectory modeling, and thermal protection analysis.

```
import numpy as np

def calculate_altitude(v0, theta, g=9.81):
    '''
    Calculate the maximum altitude of the sounding rocket.
    :param v0: Initial velocity (m/s).
    :param theta: Launch angle (degrees).
    :param g: Gravitational acceleration (m/s^2).
    :return: Maximum altitude (m).
    '''
    theta_rad = np.radians(theta)
    return (v0**2 * (np.sin(theta_rad))**2) / (2 * g)

def calculate_drag(C_D, rho, A, v):
    '''
    Calculate the aerodynamic drag force.
    :param C_D: Drag coefficient.
    :param rho: Air density (kg/m^3).
    :param A: Reference area (m^2).
    :param v: Velocity (m/s).
    :return: Drag force (N).
    '''
    return 0.5 * C_D * rho * A * v**2

def calculate_thrust(m_dot, v_e):
    '''
    Calculate the thrust of the rocket.
    :param m_dot: Mass flow rate (kg/s).
    :param v_e: Exhaust velocity (m/s).
    :return: Thrust (N).
    '''
    return m_dot * v_e

def equation_of_motion(T, D, m, g, theta):
    '''
    Calculate velocity and angle rate of change.
    :param T: Thrust (N).
    :param D: Drag force (N).
    :param m: Mass of the rocket (kg).
    :param g: Gravitational acceleration (m/s^2).
    :param theta: Current trajectory angle (rad).
    :return: Rate of change of velocity and angle.
    '''
```

```
        dv_dt = (T - D) / m - g * np.cos(theta)
        return dv_dt

def heat_transfer_nusselt(h_c, L, k):
    '''
    Calculate the Nusselt number and convective heat transfer.
    :param h_c: Convective heat transfer coefficient (W/m^2K).
    :param L: Characteristic length (m).
    :param k: Thermal conductivity (W/mK).
    :return: Nusselt number
    '''
    return h_c * L / k

# Example inputs
v0 = 300       # m/s
theta = 85     # degrees
C_D = 0.5
rho = 1.225    # kg/m^3 at sea level
A = 0.1        # m^2
m_dot = 1.5    # kg/s
v_e = 2800     # m/s
m = 500        # kg

# Calculations
altitude = calculate_altitude(v0, theta)
drag_force = calculate_drag(C_D, rho, A, v0)
thrust = calculate_thrust(m_dot, v_e)
dv_dt = equation_of_motion(thrust, drag_force, m, 9.81,
    np.radians(theta))
nusselt_number = heat_transfer_nusselt(500, 1, 0.025)

# Outputs
print("Maximum Altitude:", altitude, "meters")
print("Drag Force:", drag_force, "Newtons")
print("Thrust:", thrust, "Newtons")
print("Velocity Change Rate:", dv_dt, "m/s^2")
print("Nusselt Number:", nusselt_number)
```

This code defines several key functions and computational steps necessary for the analysis and planning of sounding rocket missions:

- `calculate_altitude` determines the maximum altitude reached by the rocket based on initial conditions.

- `calculate_drag` computes the aerodynamic drag force experienced during ascent.

- `calculate_thrust` evaluates the thrust produced by the rocket engine.

- `equation_of_motion` models the trajectory dynamics under thrust, drag, and gravitational influences.

- `heat_transfer_nusselt` calculates the Nusselt number to assess convective heat transfer impacts.

The final block of code demonstrates these calculations with example input data, illustrating their application in sounding rocket analysis.

Chapter 65

Space Weather Effects on Rocket Launch

Introduction to Space Weather Phenomena

Space weather pertains to the environmental conditions in space influenced primarily by solar activity, including solar flares, coronal mass ejections (CMEs), and high-energy particle radiation. These phenomena can adversely affect rocket launches by inducing electrical anomalies in avionics, causing thermal variations, and affecting communication systems. Understanding these effects is crucial for ensuring the safety and success of aerospace missions.

Electromagnetic Interference and Mitigation

Solar flares can generate intense electromagnetic radiation, leading to potential interference with satellite communication and navigation systems used during rocket launch. The electromagnetic field E at a point in space can be understood using Maxwell's equations, particularly:

$$\nabla \times \mathbf{E} = -\frac{\partial \mathbf{B}}{\partial t}$$

where $\mathbf{B}$ represents the magnetic field and t is time. Shielding strategies employ metallic enclosures and components to mitigate this interference, modeled by the reflection coefficient R of a material:

$$R = \left|\frac{Z - Z_0}{Z + Z_0}\right|^2$$

where Z is the impedance of the material and Z_0 is the impedance of free space.

Radiation Effects on Electronics

High-energy particles from CMEs lead to radiation-induced single event effects and total ionizing dose accumulation in on-board computer systems. The flux Φ of incoming particles is quantified by:

$$\Phi = \frac{\mathrm{d}N}{\mathrm{d}A\,\mathrm{d}t}$$

where $\mathrm{d}N$ is the number of particles passing through a unit area $\mathrm{d}A$ in unit time $\mathrm{d}t$. The use of radiation-hardened components, which are characterized by threshold LET (linear energy transfer) values, is implemented to resist these effects.

Atmospheric Density Variations and Trajectory Impact

Space weather can modulate the density profiles of the Earth's atmosphere, impacting aerodynamic forces encountered by a rocket. Atmospheric density ρ can be expressed using a modified barometric model under perturbed conditions:

$$\rho(h) = \rho_0 \exp\left(-\frac{h}{H}\right) f(t)$$

where ρ_0 is the sea level density, h is altitude, H is the scale height, and $f(t)$ is a correction factor accounting for temporal variations due to solar influence.

Thermal and Structural Considerations

Increased solar activity leads to heightened infrared and ultraviolet radiation, influencing thermal loads on rocket surfaces. The net heat flux $\dot{q}$ is described by:

$$\dot{q} = \sigma \left(T_{\text{surface}}^4 - T_{\text{ambient}}^4 \right)$$

where σ is the Stefan-Boltzmann constant, T_{surface} is the surface temperature, and T_{ambient} represents the ambient temperature. Proper thermal protection systems are designed with materials exhibiting high emissivity and low conductivity.

1 Material Selection and Stress Analysis

Materials selected for rocket construction are evaluated for their structural integrity under varying thermal and radiation conditions. Stress analysis is performed using finite element methods (FEM), solving the differential equations governing elasticity:

$$\nabla \cdot \sigma + \mathbf{f} = \mathbf{0}$$

where σ is the stress tensor and $\mathbf{f}$ signifies external forces.

Mitigation Strategies for Launch Environment

Predictive models utilizing data from solar observatories and space weather monitoring satellites are employed to anticipate adverse conditions and adjust launch windows accordingly. Algorithms for adaptive control are integrated into launch systems, utilizing real-time environmental data to dynamically adjust the trajectory and system responses, ensuring mission safety.

Python Code Snippet

Below is a Python code snippet that provides computations related to space weather effects on rocket launch, including electromagnetic interference mitigation, radiation effects on electronics, atmospheric density variation, and thermal load management.

```python
import numpy as np
from scipy.constants import sigma

def electromagnetic_interference_shielding(Z, Z0):
    '''
    Calculate reflection coefficient for shielding against
    ↪ electromagnetic interference.
    :param Z: Impedance of the shielding material.
    :param Z0: Impedance of free space.
    :return: Reflection coefficient.
    '''
    R = np.abs((Z - Z0) / (Z + Z0)) ** 2
    return R

def radiation_flux(dN, dA, dt):
    '''
    Calculate the particle flux for radiation impact analysis.
    :param dN: Number of particles.
    :param dA: Area (m^2).
    :param dt: Time (s).
    :return: Particle flux.
    '''
    Phi = dN / (dA * dt)
    return Phi

def atmospheric_density_variation(rho0, h, H, f_t):
    '''
    Calculate atmospheric density under solar influence.
    :param rho0: Sea level density (kg/m^3).
    :param h: Altitude (meters).
    :param H: Scale height (meters).
    :param f_t: Temporal correction factor.
    :return: Atmospheric density at altitude h.
    '''
    rho = rho0 * np.exp(-h / H) * f_t
    return rho

def heat_flux(T_surface, T_ambient):
    '''
    Calculate net heat flux on rocket surface.
    :param T_surface: Surface temperature (Kelvin).
    :param T_ambient: Ambient temperature (Kelvin).
    :return: Heat flux (W/m^2).
    '''
    q_dot = sigma * (T_surface**4 - T_ambient**4)
    return q_dot

def structural_stress_analysis(stress_tensor, external_forces):
    '''
    Evaluate structural integrity using finite element analysis.
    :param stress_tensor: Stress tensor.
    :param external_forces: External forces vector.
```

```
:return: Equilibrium equation result.
'''
equilibrium = np.dot(stress_tensor, external_forces)
return equilibrium

def run_simulation():
    # Example parameters
    Z, Z0 = 377, 377   # Impedance of free space (Ohms)
    dN, dA, dt = 1e8, 1.0, 1.0   # Number of particles, area (m^2),
    ↪ time (s)
    rho0, h, H, f_t = 1.225, 10000, 8500, 1.0   # kg/m^3, meters,
    ↪ correction factor
    T_surface, T_ambient = 300.0, 250.0   # Kelvin
    stress_tensor = np.array([[1, 0, 0], [0, 1, 0], [0, 0, 1]])
    external_forces = np.array([0, 0, -9.81])   # N

    # Compute values
    R = electromagnetic_interference_shielding(Z, Z0)
    Phi = radiation_flux(dN, dA, dt)
    rho = atmospheric_density_variation(rho0, h, H, f_t)
    q_dot = heat_flux(T_surface, T_ambient)
    equilibrium = structural_stress_analysis(stress_tensor,
    ↪ external_forces)

    # Summary of computed values
    print("Reflection Coefficient for EMI Shielding:", R)
    print("Radiation Flux:", Phi, "particles/m^2/s")
    print("Atmospheric Density at Altitude:", rho, "kg/m^3")
    print("Net Heat Flux:", q_dot, "W/m^2")
    print("Structural Equilibrium Analysis Result:", equilibrium)

run_simulation()
```

The code snippet includes essential functions for modeling space weather effects and ensuring the robustness of rocket systems against these environmental challenges:

- electromagnetic_interference_shielding calculates the reflection coefficient, critical for designing effective shielding against electromagnetic disturbances.

- radiation_flux computes particle flux, allowing engineers to assess radiation impacts on electronics.

- atmospheric_density_variation models variations in atmospheric density due to solar influences.

- heat_flux determines net heat flux to predict thermal loads on rocket surfaces.

- `structural_stress_analysis` utilizes finite element methods to evaluate structural stresses under combined thermal and mechanical loads.

These computations are crucial for designing robust and adaptable rocket systems capable of withstanding the variable conditions imposed by space weather phenomena.

Chapter 66

Astrodynamics of Hypersonic Vehicles

Introduction to Hypersonic Regimes

Hypersonic vehicles operate at speeds typically greater than Mach 5, navigating through a unique set of atmospheric and orbital dynamics. This regime is characterized by significant aerodynamic heating, constrained flight envelopes, and complex control challenges. The astrodynamics of such vehicles require precise computational modeling due to the nonlinear interactions between atmospheric drag, gravitational forces, and vehicle propulsion systems.

Atmospheric Re-entry Dynamics

Hypersonic vehicles encounter extreme thermal and aerodynamic loads during atmospheric re-entry. The vehicle's trajectory optimization involves analyzing the balance between drag force F_d and gravitational force F_g. The drag force is given by:

$$F_d = \frac{1}{2} C_d \rho A v^2$$

where C_d is the drag coefficient, ρ is the atmospheric density, A is the reference area, and v is the velocity of the vehicle. The gravitational influence is modeled as:

$$F_g = m g$$

where m is the mass of the vehicle and g is the acceleration due to gravity. Trajectory modeling often uses these expressions to solve the equations of motion under the high thermal loads experienced.

Control System Response

The control systems of hypersonic vehicles must respond swiftly to rapid changes in aerodynamic conditions. The control system dynamics can be evaluated through state-space representation:

$$\dot{\mathbf{x}} = \mathbf{A}\mathbf{x} + \mathbf{B}\mathbf{u}$$

$$\mathbf{y} = \mathbf{C}\mathbf{x} + \mathbf{D}\mathbf{u}$$

where $\mathbf{x}$ is the state vector, $\mathbf{u}$ is the control input vector, and $\mathbf{A}, \mathbf{B}, \mathbf{C}, \mathbf{D}$ are the respective coefficient matrices. Feedback linearization is commonly employed to manage the non-linear characteristics of hypersonic vehicle control systems.

Thermal Protection Systems

Designing thermal protection systems (TPS) for hypersonic vehicles is critical due to the extreme heating rates. The heat transfer equation governing the heat shield design is derived from Fourier's law, expressed as:

$$q = -k\nabla T$$

where q is the heat flux, k is the thermal conductivity, and ∇T is the temperature gradient. Numerical techniques, such as finite difference methods, are utilized to solve this partial differential equation for evaluating the temperature distribution over the vehicle's surface.

Propulsion Systems Integration

Hypersonic vehicles utilize advanced propulsion systems, such as scramjets, capable of operating in the high-speed regime. The specific impulse I_{sp} is crucial for assessing the propulsion system's efficiency, given by:

$$I_{sp} = \frac{u_e}{g_0}$$

where u_e is the effective exhaust velocity and g_0 is the standard gravity. Engine integration must address the complexities of high enthalpy flows and shock-wave interactions.

Navigation in Hypersonic Flight

Navigation systems in hypersonic vehicles rely on precise inertial and satellite-based methods to maintain the accuracy required for target acquisition and strategic maneuvers. The Kalman filter is extensively employed for state estimation, with its recursive nature described as:

$$\mathbf{x}_{k|k} = \mathbf{x}_{k|k-1} + \mathbf{K}_k(\mathbf{z}_k - \mathbf{H}\mathbf{x}_{k|k-1})$$

where $\mathbf{x}_{k|k}$ is the updated state estimate, $\mathbf{K}_k$ is the Kalman gain, and $\mathbf{z}_k$ is the measurement vector.

Mission Design Considerations

The strategic application of hypersonic vehicles necessitates a comprehensive understanding of mission design that integrates trajectory optimization, energy management, and sensor payload performance. Mission analysis involves computing the performance trade-offs to satisfy operational constraints and mission objectives.

Python Code Snippet

Below is a Python code snippet that encompasses the core computational elements of hypersonic vehicle astrodynamics, including the computation of drag and gravitational forces, control system modeling, thermal protection system analysis, propulsion system efficiency, and navigation algorithms.

```python
import numpy as np

def calc_drag_force(C_d, rho, A, v):
    '''
    Calculate the drag force on a hypersonic vehicle.
```

```
    :param C_d: Drag coefficient.
    :param rho: Atmospheric density.
    :param A: Reference area.
    :param v: Velocity of the vehicle.
    :return: Drag force value.
    '''
    return 0.5 * C_d * rho * A * v ** 2

def calc_grav_force(m, g):
    '''
    Calculate the gravitational force on a hypersonic vehicle.
    :param m: Mass of the vehicle.
    :param g: Acceleration due to gravity.
    :return: Gravitational force value.
    '''
    return m * g

def state_space_model(A, B, C, D, x, u):
    '''
    Compute the state-space model for vehicle control dynamics.
    :param A: State transition matrix.
    :param B: Control input matrix.
    :param C: Output matrix.
    :param D: Feedforward matrix.
    :param x: State vector.
    :param u: Control input vector.
    :return: New state and output vectors.
    '''
    x_dot = A @ x + B @ u
    y = C @ x + D @ u
    return x_dot, y

def calc_heat_flux(k, grad_T):
    '''
    Calculate heat flux using Fourier's law.
    :param k: Thermal conductivity.
    :param grad_T: Temperature gradient.
    :return: Heat flux.
    '''
    return -k * grad_T

def calc_specific_impulse(u_e, g_0):
    '''
    Calculate the specific impulse of the propulsion system.
    :param u_e: Effective exhaust velocity.
    :param g_0: Standard gravity.
    :return: Specific impulse value.
    '''
    return u_e / g_0

def kalman_filter_update(x_k_k1, z_k, K_k, H):
    '''
    Perform Kalman filter state update.
```

```python
    :param x_k_k1: Predicted state estimate.
    :param z_k: Measurement vector.
    :param K_k: Kalman gain.
    :param H: Measurement matrix.
    :return: Updated state estimate.
    '''
    return x_k_k1 + K_k @ (z_k - H @ x_k_k1)

# Example Parameters
C_d = 0.5
rho = 0.02   # kg/m^3
A = 2.0      # m^2
v = 1500     # m/s
m = 10000    # kg
g = 9.81     # m/s^2
k = 0.02     # W/m-K
grad_T = 500 # K/m
u_e = 4000   # m/s
g_0 = 9.81   # m/s^2

# State Space Example
A = np.array([[0, 1], [-0.1, -0.2]])
B = np.array([[0], [1]])
C = np.array([[1, 0]])
D = np.array([[0]])
x = np.array([[0], [0]])
u = np.array([[1]])

# Kalman Filter Example
x_k_k1 = np.array([0.0, 0.0])
z_k = np.array([1.0])
K_k = np.array([[0.1], [0.1]])
H = np.array([[1, 0]])

# Calculations
drag_force = calc_drag_force(C_d, rho, A, v)
grav_force = calc_grav_force(m, g)
x_dot, y = state_space_model(A, B, C, D, x, u)
heat_flux = calc_heat_flux(k, grad_T)
specific_impulse = calc_specific_impulse(u_e, g_0)
x_k_k = kalman_filter_update(x_k_k1, z_k, K_k, H)

print("Drag Force:", drag_force, "N")
print("Gravitational Force:", grav_force, "N")
print("State Derivative:", x_dot)
print("Output:", y)
print("Heat Flux:", heat_flux, "W/m2")
print("Specific Impulse:", specific_impulse, "s")
print("Updated State Estimate:", x_k_k)
```

This code defines several key functions necessary for the implementation of hypersonic vehicle dynamics and controls:

- `calc_drag_force` function computes the aerodynamic drag force based on current vehicle parameters.

- `calc_grav_force` calculates the gravitational force acting on the vehicle.

- `state_space_model` models the control system dynamics using a state-space representation.

- `calc_heat_flux` assesses the heat flux through the vehicle's thermal protection system.

- `calc_specific_impulse` determines the efficiency of the propulsion system.

- `kalman_filter_update` demonstrates a Kalman filter state update for navigation purposes.

The final block of code provides examples of computing these elements with example parameters.

Printed in Great Britain
by Amazon